ESSAYS AND ADDRESSES.

VOL. III.

EVOLUTION AND OCCULTISM.

By

ANNIE BESANT.

London:

The Theosophical Publishing Society,

161. New Bond Street. W.

Madras: The Theosophist Office.

1913.

Publishers' Preface.

IN addition to the large number of volumes which stand in the name of Annie Besant in the catalogue of the British Museum, there is a great quantity of literature, for which she is responsible, that has appeared in more fugitive form as articles, pamphlets and published lectures, issued not only in Great Britain but in America, India and Australia. Much of this work is of great interest, but is quite out of reach of the general reader as it is no longer in print, and inquiries for many such items have frequently to be answered in the negative. Under these circumstances the T.P.S. decided to issue an edition of Mrs. Besant's collected writings under the title ESSAYS AND ADDRESSES. It was originally intended to arrange the matter in chronological order, commencing with the writer's first introduction to Theosophy as reviewer of Mme. Blavatsky's *Secret Doctrine*, but several considerations determined the abandonment of this plan in favour of the scheme now adopted, which is the classification of subject-matter independent of chronological order. The Publishers feel sure that this arrangement will especially commend itself to students who desire to know what the Author has written on various important aspects of Theosophy in its several ramifications, and for all purposes of study and reference the plan chosen should more effectively serve. The dates and sources of articles are given in nearly all cases, and they are printed without any revision beyond the correction of obvious typographical errors.

The importance and interest of such a collection of essays, both as supplementing treatment of many of the topics in larger works and as affording expression of the Author's views on many subjects not otherwise dealt with, will be obvious, and it only remains to express the Publishers' hope that the convenience and moderate cost of the series may insure its thorough circulation among the wide range of Mrs. Besant's readers.

T.P.S.

London. May. 1913.

CONTENTS.

The Birth and Evolution of the Soul.

Two Lectures given at 19, Avenue Road, London. in 1895.

Part I.

THESE two lectures might better perhaps be described as one lecture in two parts, for I am really going to try and give you in the two a connected tracing of the progress of the soul. There is so much confusion in thought as to the origin of the individual, as to what the individual really means, as to how he is developed, and what is to be his ultimate destiny, that I thought I could take no better subject for a Lodge, which ought to be a Lodge of students, than to trace out somewhat in detail this most important matter in the light of Theosophy. Important, because on it turns your whole view of the purpose of the Universe; and if the growth of the Soul were better understood than it is, we should not hear the continual questions asked as to why there should be a Universe at all, and why there should be manifestation ; why, if everything comes out of One and goes back to One again, why this intermediate condition of multiplicity should occur. The whole of these questions really turn on misunderstanding, or on lack of accurate knowledge, and it is to the clearing up of that misunderstanding that I am going to address myself to-night and this night week.

In order to make the origin of the individual clear, I must ask you to come back with me to the time when the human race was evolving, but when, as yet, the individual had not come into existence. Come back to what you know, in Theosophical literature, as the first Race of man. You may remember that – according to the teachings which you find in *The Secret Doctrine* and in the great Scriptures of the world - you may remember that the first Race had its bodies built up, as it were, round a form that is said to have been derived from the Moon. The Pitris, or ancestors, who afforded the first matrix for humanity, who gave the first ethereal form through and by the help of which the physical in man was to evolve – those are spoken of as the Lunar Pitris, because of their connection with the Moon. Into that connection I have not time to go in detail, but the traces of it are around you on every side at the present time. You must be perfectly well aware that the effect of the Moon upon the earth is marked and constant, and above all you may notice that it is the measure in time of all great physiological periods. The fact that you still find that to be the case after so many millions of years should make it at least not surprising to you when you read in ancient books that there is a causal relation between the Moon and the Earth, and between the ancestors of humanity and forms of living beings who existed on the lunar globe in ages gone by.

Now these Lunar Pitris that came and projected their astral forms, as it is said, in order that the first Race of man might develop, came to the Earth, this Earth on which we are now, which is the central globe in the ring of globes. and they came to it - so far as we need trouble about it to-night – for the Fourth Round ; of course not for the first time, but I am ignoring the first three Rounds. When they arrived at this globe for the Fourth Round, they threw out these shadows – ethereal bodies, as they are called. The reason – and I may as well mention it in passing - why an ethereal body, or body of astral matter, should in point of time come before the physical body that we know, is that you cannot have any vital energies at work, you cannot have any electrical or magnetic currents, or any of those currents which have to do with the various phenomena of life and of chemical action, without the presence of ether. So that it is not possible to draw together physical molecules for the making of a dense body unless you have what we may call a matrix of ether through which these forces are able to act, and so to draw together, and hold together after they are drawn, the physical molecules. In fact, every physical molecule has its envelope of ether, and is permeated therewith, and it is by means of this ethereal envelope of the molecule that these life-forces are able to draw and hold the physical body together.

During the first and second Races of man, this physical body was built up by the action of what are called Nature-Spirits, who made this outer clothing of man, the tabernacle of flesh, as it is sometimes entitled. Out of the first Race evolved the second ; and of the second envolved the third. No break ; nothing that would be called a new creation, but definite and sequential evolution. The materials used in these bodies had been worked up in previous ages through mineral and vegetable and animal, and so had taken on, as far as their atoms were concerned, an internal differentiation, which is of enormous importance when they enter into the body of the higher animals and of man.

Man again is nearly the first being that appears on this globe at the stage of evolution that we are considering. Pass, then, from the first and second Races to the third. When the third Race was evolving, slowly and gradually through almost incalculable periods of time, the animal development took place; that is, the development of the physical and astral bodies went on during the first and second Races, and in the third that of the body of sensation, as you know it in yourselves and in the lower animals, the body which receives and translates into feeling all impressions from without. Your outside body receives contacts from the external universe, and certain parts of it are modified to answer to those contacts; these modified parts we call the sense-organs, but you are aware that the sense-organ in the body, while it answers by way of vibration to any vibration of its own particular class that comes to it from without, is not that which feels or perceives. This vibration has to be transmitted inwards in a very real sense, and not only inwards to the sense-centres, as they are called, in the brain; but inwards from this again, by way of the astral body – and this action shows itself always by the passing of electrical and magnetic currents – into this third body that I am now speaking of, the body of sensation. You may have a break between the outer impact and what we call feeling, between the outer vibrations of the ether on the retina of the eye and what we call sight. The power of sight, the power of tasting, of smelling, of hearing, of feeling, all these powers reside in this body of sensation, and that is, of course, why it has that particular name. Now, the lower animals have this body as well as we. You find it in the animals with which you are familiar around you. They feel, and not only do they feel, but they show emotions, passions, and appetites. You will see the passion of anger in some animals; you will see sex passion; you will find hunger, thirst – all these things are present in the lower animals, and we may group them all together, in order that a single phrase may describe these activities, as the "body of sensation," or, as it is often called, the "body of desire."

As you are a Lodge of students, I may venture to use a Sanskrit word, the word Kâma. This body is spoken of as the Kâma body, Kâma

Rûpa, which is only the Sanskrit for "desire-body"; and I use the name because I want in a moment to quote a phrase in which that word Kâma occurs. In the animal it is developed. Now, it is part of the truth of evolution that every stage of an evolving organism contains within itself in germ the next stage by which evolution is to proceed. Take what stage of evolution you like in the outer world, and you will always find that at any stage there is a germ which, if it proceeds, will develop into a new individual. This is characteristic of all forms of living things, and by the improvement from the germ evolution proceeds. You may take vegetable, you may take animal, you may take man, take what you like: always you will find that, if an individual is to be produced, a germ will be present which is to form part-helper of his growth. *Part* helper only; and this is always present in what is called the passive side of Nature.

You remember that if you go right back almost to the beginning of things you find what are called the "pairs of opposites," and the first appearance of these in the Second Logos is often spoken of as being Spirit-Matter, in order to give the two great poles of existence between which all organisation takes place. Two characteristics mark those two poles. One of them is active, and the other is passive ; one of them is positive, the other is negative; one of them is that which gives impulse, the other is that which gives form ; and these two are present everywhere, inseparable in Nature. Right through Nature, not only in physical Nature but through all realms of Nature, you will find this diversity ; and without the union of these two together you will find nowhere fresh growth, progress in evolution. There must always be the stimulating force, and there must be the form that develops. You may call them, if you like, male and female, as in physical Nature ; you may call them, if you like, Father and Mother ; but keep the idea plainly in mind, because on this the understanding of the origin of the individual soul depends.

Now, let me remind you of the line of evolution, which is the line of form, which comes from the Moon originally, and evolves downwards to Kâma. This, I say, is found in the animal just as it is in man, and you need to keep that in mind. In that Kâma, or desire – which builds a body with the power of sensation, the power of not only answering to outside impact but of translating it into feeling – lies the root of self-consciousness. Consciousness in germ is simply the power to respond to a vibration that comes from outside; and when to that power of response there is added feeling, then you get what we may call the "germ of mind" – not mind, but its germ, the negative side of mind in which and by means of which mind may evolve.

If for a moment you will look at the lower animals you will see a startling difference between the wild and the domesticated. You will find in the domesticated animal very much more of what you would call mind than you find in the wild animal; and for a very simple reason, that you will see as we go on. If you take the wild animal that has never come across man at all, you will find in it plenty of response to the impacts of the external world; but you will find in it comparatively little reasoning, little judgment, little linking together of inner sensation and outer object, unless the object be present, or unless some craving of the organism gives an impulse to action. Food, for instance, being within the sight of an animal, even if it is not very hungry, will cause movement towards the food; or the craving of the organism for sustenance will make it seek for food. But you do not get in the wild animal much of what you may call "ideal action," action without an impulse which comes from bodily necessities, or from the presence of an external object : that is, you have not there present much of what we know as mind, of which the lowest and earliest manifestations are the connection between an outer object and an inner sensation, and the power of recalling that connection and acting upon it without the object being present - the qualities which technically are called perception and recollection.

At this stage, then, of this single line that I have brought down thus, we have got as far as the development of Kâma, in which is the germ of mind, the stage in which, if there is to be a higher evolution, some impulse from without must be given. The passive side of Nature, brooded over by the Divine Spirit, could not by itself get any further than this stage - the development of a germ ; and when you have the body of desire present, that contains in it this germ of mind ; think of it now, if you like, as the female or mother side of Nature.

Now, for a moment leaving this altogether, take an analogy from lower Nature. Let me ask any of you who happen to be botanists – and probably all of you know enough of botany to follow the illustration – to take what is called the ovule of a plant, that which develops into a seed ; left to itself that ovule which is in the female organ will never be anything more than an ovule : it will simply wither up and perish. But it contains within it all the nourishment by means of which a new plant will grow ; it has stored up, as it were, a stock of food, if any new life, an individual, should there begin. But an individual cannot begin there simply by the action of the ovule itself ; it needs a stimulus which comes from the contents of the male organ of the plant, the pollen, and if that pollen throws out a minute cell which enters the ovule and comes into contact with the germ cell within the ovule, then there will be an interaction between these two microscopic cells, and by the union of the two a new impulse will be given, and an individual will result which will develop for a time within this ovule which has now become the seed, and after a time will show marks characteristic of its parents ; but it will separate itself from the parents and carry on life on its own account, having its own root, its own stem, and its own leaves. The starting-point of that is the junction of the two microscopic cells, differing in their nature, the one positive, impulse-giving, fertilizing, the other receptive, passive, nutrient, showing the characteristics of the two sides of Nature. Now we have here got the passive, nutrient side developed along this line which I characterized as lunar – coming from the moon. It is the side of form, and it may perhaps interest you to notice that it is the side that receives from outside, the passive side again, and that all these emotions and everything else in it are set up in answer to this impulse from without, thus showing its characteristic as the receptive or female side.

Now, in past Universes a process of development has gone on similar to that which is going on in the present world to-day ; in those past Universes minds were developed as we develop minds now, and their process of development will be clearer when you follow the process of development amongst ourselves. The minds that developed in those preceding Universes, that passed into Nirvâna, that passed out of Nirvâna again at the beginning of the present age, have many names both in *The Secret Doctrine* and in other books. Let us take the the name of "Sons of Mind" because it describes their most salient characteristic. They are sometimes spoken of under the name of Kumâras, which means "youths," sometimes they are called Solar Pitris ; but I prefer to take the name most often used in the *The*

Secret Doctrine, where of course you get it in the Sanskrit form, Mânasa Putra; we will take it in English, as "Sons of Mind." They have developed Intelligence. Now what is Intelligence? Intelligence is the result of vital activity working in a particular form of matter and developing connecting links between the external Universe and itself. It is a thing of slow growth; it is made by experience; it is evolved, it does not come into existence suddenly. Intelligence is the outcome of these repeated contacts, and of the working of life on the contacts; so that you never can get Intelligence apart from organism. You have something which may be called the Supreme Life; but it is a mistake to speak of It then as Intelligence; it is higher and deeper and sublimer than anything we know as Intelligence, and Its processes are far beyond and above everything that we call thought. Thought always consists in this linking together of the external and the internal, of making ideal links between the two, and hence images - ideas, as we call them; and Intelligence is only developed by the Supreme Life manifesting Itself, as what we, for want of a better word, are obliged to call Spirit in the English tongue – Atmâ is the familiar name in our own philosophy – by thus manifesting Itself in the subtlest form, and then gradually working through matter and thereby evolving what we call Intelligence. That is, all these connecting links that go to build mental faculties.

This process then had gone on in a past age, so far as these great Sons of Mind are concerned ; these mighty Spiritual Intelligences had accomplished what we are aiming at now. They are the successful men of past ages, who have developed into perfect men, perfect Intelligences, and now are, so to speak, co-operating in the building of a new race, co-operating in the production of a new humanity. But up to the point at which we are, they had taken no part in this evolution that had been going on – the physical side, the evolution of form. Now from These is to come a second line, from the Sons of Mind, Lords of Light, They are called sometimes Pillars of Light, and so on ; These coming down to the Earth, when the Tabernacles were ready to receive Them, came to give the necessary impulse in order that at this point of junction a new individual might arise, and afforded the active, impelling, positive energy.

You remember at the beginning of the second volume of *The Secret Doctrine*, those Stanzas called the Stanzas of Dzyan, which deal with the Evolution of Man. They have been said lately by Mr. Coleman to be purely modern productions ; but they were never found out in modern writings until Madame Blavatsky found them. But leave that to return to this. You will find it said that when these Sons of Mind came down, "from their own Rûpa they filled the Kâma." That is why I was obliged to trouble you with this word, because I wanted to quote that particular Stanza : "From their own Rûpa they filled the Kâma." Coming down to animal-man they threw part of Their own nature into him, filling the Kâma wherein the germ of sensation and feeling had been evolved, and They contributed to that the spark of intelligence. And so again in one of those same Stanzas it is said : "Some projected a spark." The more careful readers amongst you may remember it is said : "Some entered. . . . Those who entered became Arhats." Those are the great Teachers of Humanity in the earlier days of our Race – the fourth and the fifth Races, and the third and a half. The Great Teachers – Those who took this infant Humanity under Their care, and trained it, Those who absolutely entered into these bodies that were prepared, with Their highly developed Intelligences – were the mighty Adepts of the past ; They formed what were called the nurseries of Adepts for the present age ; the Great Teachers who came in order that this infant Humanity might be guarded and protected and helped in its earlier stages. With Them, so far as ordinary Humanity is concerned, we need not deal ; They entered in and took these bodies as Their vehicles. But They also, some of Them, projected the spark which fell into the kâmic receptacle : Their essence filled it. Now the individual begins where that union takes place. Before that there is no Ego in man ; before that there is no Soul in man in the full sense of the term, although the word Animal Soul is occasionally used for the feelings, emotions, and so on. The lower Soul this is often called, or the Animal Soul ; but the true Ego, that which is capable of achieving immortality, is not there. Remember how that phrase is used sometimes ; it has not necessarily immortality in itself, although it has in it the power of achieving immortality, by virtue of its connection with these immortal Sons of Mind, Who have already achieved. Man may become immortal "if he will." That was a phrase used, you may remember, in one of the letters from Master K. H. to Mr. Sinnett, published in *The Occult World*; part of the work of the Society was there said to be to teach man that he may become immortal if he will, not that he necessarily is immortal, but that he may achieve immortality. Immortal in the essence of the Soul? Yes; but not in its developed self-conscious intelligence. For intelligence has to be worked out and built up by slow degrees: intelligence has to be evolved by this spark, working through the matter into which it has come, and unless it works successfully, acquires experience slowly, and gradually builds it up into faculty in the course of that pilgrimage of the Soul that lies in front of our thought, immortality will not be achieved. For it is necessary, in order that immortality may be achieved, that this which is to acquire experience and build up accumulated experiences shall regain unity. That which is compounded does not last; that which is compounded will be at some time disintegrated; only the unit persists. The individual begins at this point, and he is a compound. He will weave into his own existence all these endless experiences, and will become so to speak, more and more compound, a more and more complex combination. But this has in itself the seed of destruction; everything that thus goes on combining has in it the conditions of disintegration, and the compound disintegrates. How, then, can this compound achieve immortality? By a process of unification that will form the last stages of its pilgrimage; by that Yoga, or union, which will make it again the One. Having achieved individuality by many, many incarnations, through which this individuality will be built up, it then unifies all these experiences, and by a subtle alchemy extracts as it were a unit experience out of the multiplicity, and in a way beyond words – beyond words because it is beyond brain-experience and thought, but which is not beyond the "sensing" of some who have at least begun the process – this individual evolves into a unity higher than its own combined nature; and while it may be said to lose individuality as we know it, it gains something which is far greater. Without losing the essence of individuality it re-becomes a unit consciousness, and by that becomes incapable of disintegration and achieves its final immortality. But here is the beginning point – and on that I want to lay a good deal of stress – that it begins then, that before that the Ego which is now in each of you was not in existence as Ego, any more than the plant which will develop from a germ, if the germ be fertilised, is in existence before that fertilisation takes place. True, that which will form it exists, because there is no increase either of energy or of matter ; but the combination which makes the new individual does not exist until the junction has taken place, and the Ego does not exist before this union has taken place. It is there that originates the individual. You will forgive me for repeating that so often. But this is the point where the mistake comes in, and where there is so much confusion in thought ; and it is because of that that I am laying stress upon it, in order that you may have clearly in your minds this fact : that individuals begin in each Manvantara or Age, that the purpose of each Universe is the evolution of individuals, that the Universe comes into existence in order that individuals may be born, that it is maintained in existence in order that individuals may be evolved, that when it passes out of manifestation its harvest is the perfected individuals who regain unity and outlast the Universe, passing into what is called Nirvâna, to re-emerge for a new Universe as Sons of Mind, if in the former Universe they have been completely successful. There are other intermediate stages, points where failure may come in, and where evolution may have to be taken up again as it were midway, points of failure in one Universe that do not throw back the fallen, as Master K. H. pointed out, to the beginning of things again, but are such as to allow them to take up their evolution at the point where it ceased. The failures of one age become, so to speak, the pioneers of another. But leaving those complications out of consideration, the harvest of every Universe is these triumphant individuals, who have evolved unity out of diversity, and thus have achieved their immortality.

Realising that, then, let us take our individual and see what kind of an entity this is at its origin. And I think I will throw in here a very, very brief digression, which will make it a more living thing to you. Take one of the lower animals. Now we will come to the domesticated. I mentioned that with regard to the wild animal there is the germ of mind, but very little that you can really call mind. Suppose you take an animal and domesticate it, and suppose you domesticate it for generations, you will have handed on in the three bodies of that animal – the physical, the astral, and the kâmic – you will have handed on a very definite heredity; and if these individuals are domesticated time after time you will find greater and greater intelligence, as it may be called, evolving.

Now, supposing that you take a puppy, and

supposing that from that puppy's birth you keep it continually with yourselves, and you do not permit it to associate with the lower creatures, but you keep it with yourselves. Some lonely person, for instance, takes a puppy, and it is always with him or her; what is the result? The result is that in that puppy, as it grows up, there is developed a startling amount of some quality that you are forced to call Intelligence. You will develop in it a limited reason; you will develop in it a limited memory; you will develop in it a limited judgment. Now, these are qualities of the mind, not qualities of Kâma. How is it that in this lower animal these qualities are developed? They are developed artificially by the playing upon it of the human intelligence. To that animal the mind in you to some extent plays the part which the Son of Mind plays to Humanity; and thrown out from the comparatively developed Intelligence in man, these rays, these energetic rays of mental influence, vitalize the germ in the Kâma of the animal and so produce artificially, as it were, an infant mind.

Now I say that, in order that you may realise more clearly perhaps than otherwise you would, the first slow stages of the growth of mind. Let me say that this process is not good for the animal, and it is not good for the human being who does it. Neither the one nor the other is the better for the process, in fact very often both are exceedingly the worse, and it is not a wholesome practice – this over-stimulation of the domesticated animal and this artificial forcing of a mental life for which the animal body is not yet fitted, for which the animal nervous system has not yet developed the proper natural basis, and in which it is really forced, in a kind of artificial hot-house, to the detriment of the creature, and probably to its retardation in a later stage of its existence.

But it is well to remember that there is no such thing as a break in nature ; every evolution is sequential, and it is therefore possible to force evolution in this way, although it be unwise.*

Coming back from that little digression, let

* It is with much inner pleasure that I find that a statement current in Theosophical circles, and repeated by me above, is incorrect in fact. It seems, with regard to some animals at least – as the dog and the cat – that the development caused "by the playing upon it of the human intelligence" is well caused, and lifts the animal forward, so that the germinating individuality does not return to animal incarnation, but awaits elsewhere the period at which its further development shall become possible. The "forcing" is therefore helpful and beneficial, not harmful, and we may rid ourselves of the incongruous idea that, in a universe built on and permeated by Love, the out-welling of compassion and love to our younger relatives is injurious to them. There are a good many Theosophists, I think, who will share my pleasure in getting rid of a view against which one's instinct secretly rebelled.

me take up again my infant Soul, to whom I will give the name of the baby Ego, and he is very much, as regards his mental capacities, what the new-born baby is as regards his power of manifesting these faculties. Of course, the new-born baby has mental faculties which very soon force the brain to prepare itself for their manifestation, so that there is not a real analogy between the two. The want of knowledge, so to speak, in the new-born baby is simply due to the clumsiness of the instrument ; the brain is new, and it takes some little time for the links to be set up between the instrument and the player. But the player is there, when you are dealing with our race at the present time, and therefore we have not really the condition in which the Ego itself is in the state of babyhood.

Now this entity which has thus been formed at this junction of the two lines, and which I call the baby Ego, is absolutely ignorant. It has no mind, it has no thought, it has nothing more than the sensation it gets from Kâma at present, except the power of evolving which it has received from the stimulating spark of the Son of Mind. Sensations are there ; it has to make the link which we call perception. How will that link be made? The sensation will be caused by way of the body through which it has come into contact with some external object. Let us say that the body, by the mouth, comes upon an external object which gives rise to a pleasant sensation of taste – something which is sweet. The animal of course has developed this already, and in the body it will be a habit that when it sees this thing, or feels hunger, it will go towards it. The baby Ego will experience the sensation which is pleasurable, but it will only be a momentary sensation, and at first apparently nothing more – a little impact on this germ of mind ; over and over again such an impact will take place. At last there is set up in this baby Ego by this repetition a connecting link between the external object that gives rise to a pleasant sensation and the pleasant sensation, and it will thus make its first *thought*. This connecting link between the external and the internal, between the contact which comes from an outside object and the pleasure which that contact gives, will be what is called a " percept," and you have in perception the first activity of the mind ; when this perception has been repeated over and over and over and over again, it will be remembered, and conscious memory begins.

Built up in this baby by these repeated contacts, and repeated pleasurable sensations, and repeated connections between the object and the sensation, at last memory will develop, which is the ideal contact ; the idea is built out of a number of these sensations. And that faculty of memory will be a faculty of the baby Ego which will be evolved by these constant experiences ; and it will take a long time evolving – perhaps a whole life, or part of a life. I cannot measure it off ; but I want you to realise that it is a thing which will take a considerable time, that the memory will be a thing which will need much experience at this early stage, before it will really become recognisable and workable.

There are human beings even at the present time in which this faculty of the Soul is so little developed, that it will not last over even twelve hours, and in which their view of the world is quite different in the morning and in the evening. Some of the lowest aborigines in Australia in whom the spark has burned very low, in whom it has not developed and grown, are on record as having so little memory that they cannot remember through the course of a whole day, and blankets given away in the evening will be clung to because the night has begun and the night is cold. But when the next morning comes round and the immediate use for the blanket is over, and while they do not want the blanket they do want food – the food is an immediate want, but the blanket won't be wanted till evening – some of them have not sufficiently developed what we should call the idea of invariable sequence in their minds to remember that night will come back again after the day is over, and that the blanket which they do not want while the sun is out they will want when the sun has set. The sunset of yesterday is, so to speak, a past incarnation to them, and they do not carry on the memory through the night; therefore, they will part with their blanket for a mere trifle in the morning, although they will not part with it in the evening – a most striking illustration of the baby sense, if I may so call it, of the Ego which is incarnated in these aborigines. They are dying off very fast, and no English government will be able to keep them alive, because their work is done. A race dies when it is of no more use to the Soul; it becomes sterile when its purpose in the evolution of the Soul is over. For as the Universe only exists for the sake of the Soul, so all these stages in the Universe exist for the Soul, and when there are no more Souls so little developed that such a race is of any use to them, it becomes sterile and gradually disappears. I do not mean when it is helped to disappear by the superior races, though that is often the case; but even if they do not help it to disappear rapidly, it will inevitably disappear slowly on its own account, by the barrenness which falls upon it. It is of no more use, therefore it does not continue.

That illustration may give you some idea of what the spark is like in its early stages, as you find it in these sparks that have burned low and not developed. Memory will be very, very slowly developed, but when it is developed, even in a limited way, you will at once see that an element is present conducing to more rapid growth, because the moment that this baby remembers past experiences it is then beginning to accumulate a little store which will impel it to action without impulse from without. It will have an impulse beginning from within which will lead it to seek experiences. Farther, not until it has memory can it distinguish, in their absence, between pleasurable and painful experiences, good and evil, as it will call them, and so begin to develop in itself a power of comparison and selection, *i.e.*, of judgment, which will serve as a guide for action. Let me take the case of taste, which I chose before. It was pleasurable. Of course, some tastes will be unpleasant, and those will be marked off as painful, to be avoided. So that the Ego will get, as it begins to remember, two classes of things in the outer world ; one labelled in its own mind – if I may call it mind at this stage : "pleasure ; to be sought, to be followed ;" and the other labelled : "painful ; to be avoided, to be run away from, to be escaped." And at this stage, the Universe, so to speak, will divide itself into two for this baby Ego, things to be run after and things to be run away from. It will not have gone any further than that. Going out into an unknown world where it comes into contact with objects, the first great division will be things that it wants, and thing that it does not want ; and the wanting or the not wanting will depend on whether it meets something which it desires to repeat because it gives it pleasure, or something which it desires to avoid repeating because it gives it pain. This is the beginning of experience.

In this way it will begin, as it accumulates these experiences of pleasure and of pain, to learn something more ; it will begin to learn that this world that it has come into is a very definite kind of thing, and that it has got to accommodate itself to it ; it will find some things in it that do not give way, and that if it runs up against those things certain unpleasant results always follow. Memory of course is wanted for this, to notice that always the same thing comes from the same object under the same circumstances; and when sufficient of these sensations have been accumulated to give rise to the definite idea that doing a particular thing will cause pain, *there* is the first glimpse of law, of something external to itself which it cannot overcome, which throws it back, as it were, and gives what it feels as pain, something repellent when it comes against it. So the idea begins to arise that not only are some things to be followed and others to be avoided, but that these things which are to be followed, and which are pleasure-giving, are things which are "good" – which only means at first harmonious – and that others are inharmonious and unpleasant, and therefore "evil"; that there is a law of pleasure and of pain to which it must adapt itself if it wants to live in comfort, that nothing that it can do will break this law, and that, therefore, it will be wise to accommodate itself to the law. This observation of sequence will be made by our baby Ego and will give rise to the idea of Law, and of the need to adapt itself to these laws if it is to live at all comfortably. And then a little more will come as the experience goes on ; that sometimes a thing begins by giving pleasure and goes on by giving pain - a most confusing experience. Let us cling to our taste. The body of our baby Ego eats and pleasure is felt; because of this it makes the body go on eating till it eats too much. It then finds that by repeating this gratification pain has come where there was pleasure. It makes its body ill, and it gets a new view of the outside Universe - that the gratification of this which began by being pleasurable works out into pain, and that the pleasure which began in flavour may end in most uncomfortable aches ; and not only so, but, persisted in, may cause perennial aches which later on it will know as disease. This very much emphasises its idea of law, and it begins to accumulate now a sequential experience of different pleasures and pains, and to realise that it bears a certain relationship to these outside contacts ; it learns that it is not the outside object in itself that is pleasure-giving or pain-giving, but some relation that arises between itself and the outside object – a great advance – and that these outside objects are neither pleasure-giving nor pain-giving in themselves, but only in relation to itself, the same thing sometimes giving the one and sometimes giving the other. So that the idea of pleasure and pain, in this further experience of our baby Ego, will go on into the relationships that itself sets up with the outer world, and that change the character of the outside impact from pleasurable to painful. And then the law will begin to take on, as it were, a compelling power, and it will realise that it can adapt itself to this strange external apparent change, and that by adapting itself it can persist in pleasure or persist in pain, and that the pleasure and pain will depend on its attitude to the outer world. And so this next lesson of experience will be learned. And there I must leave my baby Ego for to-night, having reached as far as the recognition of an outer world, the receiving of pleasure and of pain, the recognition of relations, therefrom evolution of memory, evolution of judgment – which recognises the relationship as having in itself this difference of pleasure and pain – so that we have the beginnings of perception, memory, judgment – three things that are wanting for what we call reason of an elementary kind. Reason only exists certainly in the baby Ego as a mere germ ; and we will leave him as he passes through death, carrying with him these germinal mental faculties which he has evolved. We cannot say how much progress would be made in one of these early lives ; probably many lives would be needed to arrive at the stage just described. In order to make our study complete in outline, let us take him at the end of his first life, to see the principle underlying *post-mortem* evolution. Let us see him having begun his pilgrimage and passing for the first time through the gate of death. On the other side of that we will leave him to take him up again next week.

PART II.

YOU will remember that last week we left what I called the baby Ego having passed through the gateway of death. Now I spoke of his passing for the first time through the gateway of death, because I wanted to take up for a few moments the *post-mortem* states in order that we might have as it were before us for the rest of the study of the pilgrimage this complete cycle : the life upon earth, the life in the transition state beyond death, the life in the Devachanic state - the life of the Soul properly so-called, the intellectual and intelligent life.

Those three stages, being the three divisions of the pilgrimage completing a single period, are repeated over and over again, succeeding each other in this definite way, all bearing to each other a definite relation, so that unless we understand each of the three we shall not be able to follow with any intelligence the pilgrimage of the Soul or the growth of the Ego.

When, however, this germinal Ego that I spoke of last week passes for the first time through the gateway of death, it has exceedingly little material for these stages that lie on the other side of that gateway. The first stage is that which may be called – translating the term – the Land of Desire. You will remember that desire is Kâma, and Loka is place; so that this land or place of desire is called, in Theosophical books, Kâma Loka; that is, a place inhabited by Souls still clad in the desire or sensation body that we studied last week. The Ego in this body dwells there for a time – but not only the Ego of man. It is the place where these bodies of sensation and desire survive the physical and the astral bodies; so that you have there these desire or sensation bodies of the lower animals as well as those which are inhabited by the Human Soul.

Now, when our baby Ego is at this very

early stage of his life there will be in him a great deal of the lower element of Kâma or desire, and scarcely anything at all of the higher element of Mind. His stay then in this Kâma Loka will be for a considerable period, and all that he will do there is to experience pleasant or painful results according as the life which has been led on the physical plane – which has been led upon earth – has been in accordance with law or discordant with law. The life there is exceedingly limited, and is simply a result from the life upon earth. Nothing new is introduced into it ; it is a life in which there is a great deal of repetition, in which an experience is repeated over and over and over again. And this automatic action, as we may call it, is one of the great characteristics of this Kâmic body, or body of sensation. You know how easily habits are set up – habits of the physical body, especially habits which are connected with the passions and with the emotions. The great root of these habits lies in the body of sensation with its peculiarly strong automatic tendencies. Those are impressed on the outer body, although, of course, the habit of repetition in the physical body will also come in to some extent.

When the Soul has passed out of this transitional state it leaves behind it this body of desire, just as in leaving the earth it left the physical body. The bodies belong to certain definite stages in the Universe, and the Soul cannot carry on any body with which it has been clothed further than the stage in the Universe to which that body belongs. It cannot carry the physical body away from the physical plane ; it cannot carry the astral body out of the lower astral world ; it cannot carry this body of sensation out of the transitional state, known as the land of sensation or desire. And when it has worn it out sufficiently to allow it to escape, when it has exhausted, so to speak, this body of desire, which has been nourished in the sensational life of earth, then it passes on into a higher condition, into a higher sphere, where the whole of its work is the work of the mînd, all higher aspirations, all thoughts which are devoid of passion and of appetite, everything which is intellectual as distinguished from what is passional, all the higher emotions which have in them this element of mind as well as the lower element of passion – these, purified from passion, will be carried on into this higher world. And the length of the stay of the Soul in that world depends on the amount which during its earth-life it has accumulated of mental experiences, and of experiences of the higher emotions, of the artistic faculties, and so on, everything in fact which has to do with the mind. Understanding that, you will see at once that when the Soul first passes through the gateway of death there will be scarcely anything for it to carry on into this higher condition, hardly any experiences which it can use for the development, as it were, of mental faculty. Still, the very few experiences that it has acquired during its first life in the body, which are not kâmic, will be carried on. What will be its work then in this higher world? It will be to extract from these separated mental experiences their essence, and to transmute that essence by working upon it with this energy of the Soul, transmute that essence into mental faculty, or mental ability. The work that the Soul accomplishes when it is out of the body, when it can no longer gather fresh experiences, when it has lost the three bodies through which experiences can alone be collected, the work of the Soul is then to take up the mental images remaining from these experiences, and, working on them, to take out of them their essence. Just in the same way that a chemist might take a number of chemical elements, might throw them into a crucible, and then purifying them from dross might extract the elements themselves and combine them in the crucible; so does this chemist, as it were – the Soul – by the alchemy of its own mental ability, the thought power which it has developed, working on these accumulated and separated experiences, throwing them into the crucible of the mind, extract from them their essence, and then taking that essence it assimilates it, makes it part of itself, works it into its own nature, or – to use the phrase that I used two or three times last week – weaves these separated experiences into itself, and so begins to make a real garment of the Soul, which is the character of the Soul, and which will reappear as character when it comes back to earth. Everything that the Soul brings with it of mental faculty, everything that is born, as we say, with the child, the powers of the mind which the child shows – the whole of these are brought back by the Soul as the result of its workings on past experiences while it is living in the world of the Soul, the world that we call Devachan in our Theosophical literature, which simply means the land of bliss. As I say then, in these early experiences there will be very little for the Soul to work upon in this blissful land; but when it comes back even after this first period is completed, and comes to be born for the second time upon earth, it will have what it did not have at the beginning – a little germ of mental faculty. That is the small result in faculty which it has brought back by working on the few experiences that it accumulated during its previous life. It will start then at a certain advantage in this second period of its pilgrimage; it will start with certain tools, as it were, ready made to its hand, which it has fashioned for itself during this interlude in the world of the Soul; it will come back with a nascent memory, with a nascent power of comparison, very, very small certainly, but still, so to speak, better than none; and it will work through that on the new experiences that come to it by way of sensation from the outer world.

Through this life, then, again it will pass, having this advantage now, that it has a little mental faculty to go upon which it can increase. As it goes on experiences increase, its power to receive the experiences being greater, and this mental element mixing itself up with the emotional and the passional nature. So that we now speak of the mind, or as we call it, Manas – the word from which our own word man is derived, and which really means the thinker: the great characteristic of the man being that he thinks. This Manas, then, now coming back, small as its powers are, will modify and change the whole of the kâmic nature, the passional nature ; and all that this Ego now experiences will have in it the two elements – the element of passion which belongs to its passional nature, and the element which comes from this mind which is developing, which tends gradually more and more to observe, and to compare, and to make record of its experiences, and to store them up in order that it may direct its action by them. At first all the actions will grow from outer attraction ; presently against the outer attraction there will be working the images of past experiences. So that, to take up the illustration that I used last week of taste, when there is a strong attraction from without of a taste which it knows will be pleasurable, excess will be guarded against by the mental image which has been preserved of the pains that in previous experiences were the result of over-gratification of taste. So that now you will have this double element. And remember that the element of the mind is increasing, while the other element is more or less stationary. As the Soul passes from life to life, and in each life accumulates experiences, in each transitional stage after death suffers from the animal appetites which hold it prisoner from going on into the happier world ; and then in that happier world works upon the experiences and changes them into faculties, it will always have an accumulating stock of faculty, an accumulating store of memories, while the outer attractions will remain comparatively the same, and action will be more and more directed by reason and less and less directed by appetite.

Now, understanding that you will be fairly able, I think, to trace, so to speak, the stages of this pilgrimage of the Soul : you see the elements that enter into the pilgrimage ; you see the tools with which the Soul will have to work improvements, if it uses its experiences well in the successive incarnations ; and you will also understand that on the accumulation of these experiences, and the working upon those experiences in the blissful land, will depend the more rapid or the slower growth of the faculties of the Soul – that is, whether it will grow rapidly, or will grow slowly, or moderately, whether the pilgrimage shall be comparatively swift or very much delayed. You will see also how the Soul may often be thrown back, how a very strong attraction from without may overbear, say, a comparatively small store of accumulated experiences ; and then the Soul will make a mistake, will go against the law, and will suffer.

How should we regard such an experience when we are studying this pilgrimage of the Soul? Is it a matter for very great regret? Is it a matter for extreme grief and sorrow?

Think it out for yourselves and you will see the way in which this wider view of life will regard any mistake, any blunder, any fall, any sin. Sin is disharmony with law. So long as the law is not understood, desire will constantly be drawing the Soul outward without regard to this law of which it knows nothing, and striking on the law it will feel pain. Suppose then that it has not accumulated a sufficient store of these painful experiences to make it realize the presence of the law; or suppose that having accumulated sufficient experience to recognize the presence of the law, it has not accumulated sufficient to overbear the strong drawing of attraction to the external object; then the very experience of wrong-doing is a necessary stage in its education. For the pain that results from the wrong-doing will add to the store of experience that the Soul is gradually accumulating, and will make it stronger against the temptation in the future by this new suffering which it has found must inevitably result from coming into conflict with the law. So that instead of being heart-broken over a failure, those who see from life to life and look on the pilgrimage of the Soul as a great whole, and not simply in the fragments that ordinary persons see in looking at it, they can see with calmness these blunders that the Soul makes, knowing that they are the result of insufficient experience, and that the very fall will supply an added experience which will help the Soul to stand when it comes into a similar position in the future. And there is no more reason for extreme sorrow over these blunders of the growing Soul than, to use a simile that I have often used before, there is reason for the mother to break her heart because the child may stumble when it is learning to walk. If the child is hurt she may feel sorrow for the child's pain, but she certainly will not go into a state of frantic despair about it. She will know that these tumbles are a necessary part of the education of the child in gaining equilibrium, and will know that every tumble it has will make the tumbles of the future less likely to occur.

Now, that is not a callous way of looking at things ; it is a wise way; and as knowledge grows wisdom gives balance to contemplate calmly many things that otherwise would be distressing and disturbing in the very highest degree. Calmness, which is a characteristic of wisdom, comes from this wider vision which is able to understand causes as well as see effects, and which understands how that which to-day seems sad will work out for good in days to come.

One other distinction that we want to realize as to the principles at work in this pilgrimage of the Soul, is that the kâmic element with which the Ego is encircled, constantly giving rise to desire, is that which is always making links which bring it back to birth. Every desire that you have for something down here survives death, remains behind you in the transition state until you return for reincarnation, and draws you back to rebirth as soon as you have exhausted in the blissful land the accumulated experiences of the mind which you work upon in that region of the Universe. So to speak, when the body drops from it, the Ego, having accumulated sufficient materials for its work, has the tendency to leave the Earth and assimilate what it has gained through these agencies which exist in the desire body. It leaves earth behind, being drawn by the stronger mental forces onward, where it has to work in the region of ideas. When its store is exhausted, then the desire links re-assert their power, and the Ego having finished, having got through all the mental experiences in the blissful land, feels again these links of desire reasserting their power, and it is drawn back by them to re-incarnation, and attached by those links of desire to all the objects of desire, with which they are connected.

Now the especial reason that I mention this is that I hear so many people when they are dealing with reincarnation say, " I don't want to come back." It is of no use having a theory that you do not want to come back, if you are making these desire links to things on the physical and on the kâmic planes. So long as you want anything which the world can give you, your Ego *does* want to come back, and must come back whether you think you want it or not. The fact that it desires something here shows this fundamental craving for return, and the mere feeling of weariness, which is the outcome of a tired brain, and of a consciousness working in that tired brain, has absolutely nothing to do with the inevitable destiny of the Ego. The brain which is tired certainly will not come back; it will go to pieces on the earth to which it belongs, and the tiredness which makes people say, " I don't want to come back," is the tiredness of this outside body, the desire to escape from the painful things that have made an impression upon it, and so on. The real desire is shown by the attachment to the things of earth, by the wish for one thing or another, the wish for ease, the wish for pleasure, the wish for social consideration, the wish for the praise of men, the wish for everything with which men's and women's lives are filled well-nigh to the brim at the present time. For persons who are full of desire in this way, to say: "I don't want to come back," is simply to show a lack of understanding. They must come back until there is nothing here which has the slightest attraction for them. When nothing here attracts them, when praise and blame are exactly the same to them, when they do not mind whether in the outside world they are rich or poor, when they do not mind whether they are what people call happy or unhappy, when the whole outside life is an absolute matter of indifference, when nothing can shake their peace or bring the slightest ruffle of any kind over the emotional nature, then that Soul is ready to go on; but so long as any of these things have the slightest influence, so long these links are being made and must draw the Soul back to fresh experiences of earth.

Manas itself does not make these links, it

makes them through Kâma, and it makes them where the desire even for intellectual things comes in, but not by pure abstract thinking, which is its own special work. That is to say, it is not Manas pure and simple which makes links bringing us back to rebirth, it is Kâma-Manas, which is the form of Manas working amongst the great masses of people to-day. Manas itself, which comes out now and again in absolutely abstract thinking, does not make links which draw to rebirth. But inasmuch as almost all intellectual effort here is very largely carried on with the kâmic element, and has worked through the brain, which is the vehicle of Kâma-Manas, most intellectual effort will have in it the desire element, and so will bring the Soul back for fresh experiences upon the earth.

As to the way in which it works : it works by what we may call the creative power of thought. The whole world is the outcome of the Divine Thought. Everything which we know as phenomenal is the mere outside appearance which has in it the inner and living reality of thought. All outside appearance is but the form which the thought takes for expression on these lower planes ; and the whole Universe is nothing more than a Divine Thought in manifestation. That Divine, that God-like element is in man, and it works through Manas, and is the creative element. The more of that there is in the activity of a person, the more is he a creative energy in the world.

Every thought makes to itself a form. Every time that you think, a form is made in your mental atmosphere. A passing thought will only have a very transitory form, a thought which is constantly repeated will have a form which by these repetitions becomes stronger and stronger, and more and more permanent, so that according to the fixity and the motives of your thoughts will be the life of the thought-forms that you are continually generating around you.

If you refer to a letter in *The Occult World*, written by Master K. H. to Mr. Sinnett, you will see that He points out that when a thought goes out and takes form, it is vivified or entered into by an Elemental, and the character of the Elemental will be according to the character of the thought, and according to the motive that has inspired the thought. If the thought be a good one, for instance directed to human service with a desire to serve, then it will be helped from outside by this Elemental which is of a good and a pure type, and the thought will be a force for good, reacting on the person who has thought it, and reacting on all those who come within the sphere of his influence. So that every thought which is loving and helpful lives in the world of thought as a useful influence. And supposing that these good thoughts are directed towards people, then they go to the people to whom the will directs them and, so to speak, encircle them wîth a protective and aiding power. And it is a real thing that every good and kind thought that you have of a person, every wish for their benefit, every desire for their happiness, is an actual living thing that goes to that person as a living entity, and lives as it were in connection with the person towards whom you have directed it as a protective agency, warding off danger and drawing good towards that person to whom you have sent this angel of your thought.

So again, with all evil thoughts, thoughts which have in them the element of hate, of revenge, of passion – those draw to themselves from the outer world Elementals which increase this energy. So that an evil thought directed against a person is an absolutely mischievous agent, which may injure him either in physical health, in the astral body, or in any part of his body or mind. Suppose the person has nothing in him which in any sense forms a link with this thought of yours, then the thought will be thrown back, and will return to yourself and strike you to your own injury. Suppose, however, that the person has, what most people have, some little fault in himself, which may make a link with this thought of yours, then the evil thought attaches itself to the person and injures that person in some part of his nature. Therefore is it that everything which is of the nature of evil thought consciously directed towards a person has been called, and rightly called, Black Magic. A thought of revenge or of anger which is directed towards any person with a view to injure him is essentially of the nature of Black Magic. And the greater the power of the person who does it, and the greater the knowledge of the person who does it, the greater is their crime for which they have to answer to the Law. In this way, then, the Soul works by these thought-forms. First these thought-forms, then their Elementals, working back upon the Soul that generates them, as well as working in the outer world; these go on with the Soul into Devachan, so far as they are pure in their nature, and make the faculty of which I spoke, being as it were moulded into faculty; and then coming back, the physical body is moulded by way of the astral to manifest this character, which has taken to itself, by means of these thought-forms, certain definite shapes. So that it is perfectly true that the outer body of a person will show something of its character. That body is physically built on the aggregated thought-forms which have been transformed, by the alchemy that I spoke of, into definite faculties, and these having their characteristic forms will mould the shaping of the outer body. And it is perfectly true that when you are dealing with the general shape of the brain, you will have that brain developed in certain regions, according to the character of the Ego that inhabits it. The mistake is to suppose that it is the tabernacle which makes the tenant; it is the tenant who builds the tabernacle. So that while you have the two correlated the one to the other, we must not begin at the wrong end, and believe that the master-builder is made by his house; he builds himself the dwelling in which, in his coming incarnation, he will have to live.

One other point as regards our back-coming Ego. You will notice I am not tracing him life by life. I am giving you general principles which will work through large numbers of lives. These stages being passed through, what circumstances will our Ego come back into?

That will depend upon the circumstances that during his past life he caused upon earth; according to the happiness or misery he caused upon earth in a past incarnation, so will be the circumstances in which he will find himself when he comes back to earth.

Suppose, for instance, that a man whose influence extends over large numbers of people spreads happiness around him on every side. That is a distinct effect that he has worked, and it will govern the condition in which he will be born in his next life. This happiness that he spreads amongst large numbers of his fellows is a seed which will spring up as happy circumstances for his next incarnation. Sowing happiness he will reap happiness. Sowing pain he will reap pain. If he causes a great deal of physical suffering in his life, he will reap much physical misery in some following incarnation. If he spreads around him much mental distress and trouble he will reap mental distress and trouble in the circumstances that come in his way. Mind, these are things he cannot alter. They are fixed future events, so to speak, when he leaves the earth. These are the things that can be predicted of him with fair certainty, because these are seeds that are left which have to grow up each after its own kind.

Over these he has no power ; they are there, and he has got to live amongst them.

Now you may have a man who is not a good man morally, but who has yet spread a very large amount of happiness amongst people, say of a physical kind; he will reap physical happiness in his next incarnation. You may find a good man who by lack of knowledge has spread a great deal of misery, and he will reap physical suffering in his next incarnation. You have to distinguish, if you want to understand, between these different agencies of the Soul. According to his desires and his will, so will be his faculties – his own personal possessions, or individual possessions rather, if I may call them so ; according to what he has sown upon earth will be his harvest upon earth when he returns. So that all these circumstances of happiness or of misery will be the result of the happiness or the misery that he has spread in previous incarnations : they will come back to him as environment, as circumstances.

Now I come to my next point. You must remember the pilgrimage of the Soul is very long, and a lecture is very short, so that I am obliged to run somewhat rapidly from one subject to another. The next point at which I must make a moment's pause is on our Ego when he has become more experienced. He is no longer a baby, nor even a child, nor even a youth; he is a mature Ego, and he is becoming wise. With this wisdom of his he is bringing back more and more faculty, he is bringing back more and more memory, he will make for himself instruments which will be able to express greater and greater capacity, and a time will come in this pilgrimage of his in which his constant efforts to impress on these lower tabernacles his own ever-lengthening memory of past experiences will become more and more successful. His will having grown very strong, will tell considerably upon his lower nature. What we call the voice of conscience will begin to make itself heard with more imperative force. Now conscience is this memory of the soul expressing itself in the lower nature; it comes with authority, and the lower nature feels the authoritative sound in it: "You ought to do this, you ought not to do that." And sometimes the lower nature will challenge it, not being able to understand where this authority comes from. The authority lies in the Soul, which is trying to make the lower nature go its way; it is using its own past experience to prevent the lower nature being led astray by the outside objects, by its mistaken deductions, by its very incomplete experiences. And it is speaking constantly to this lower nature, and constantly the lower nature does not hear. In all the clatter and jangle of the body in which it is living, it finds it is very difficult to make its voice heard coming from the higher planes. But the voice of the conscience is always this voice of the Soul, speaking out of its memory. And if you think that out at your leisure you will see how it is that conscience will sometimes speak wrongly as to choice of action, but always with the sense of: "You ought to do." The reason that it sometimes speaks wrongly as to action is because the experience is still a limited experience. The reason why it is always imperative is because that limited experience is the only guide which Manas has, and it is the best guide even though it be imperfect. A man therefore does wisely always to obey his conscience. It is the best decision which experience enables the Soul to make, and if it be faulty, it is faulty because of the want of experience. If you obey it, when it blunders you will gain the lacking experience; and you will suffer more if you do not obey it. By following some other rule which is not the rule of your own inner Self, speaking from its own experience, you will be obeying an external law, which, speaking from without, is not to be relied upon to develop your Soul. The Soul is developed by experience, not by compulsion; and an outer law, however good it may be, does not, being a compulsory power, add to the inner forces of the Soul; therefore is it of comparatively small value in evolution, far less than the voice of conscience, even when the conscience is faulty.

Now taking that, let us come back from that slight digression to our Ego. It has become comparatively mature, it is getting wiser. Getting wiser it wants to escape from this constant succession of births and of deaths of which it is beginning to get a little tired. It has gone through it so often that it has accumulated a great deal of experience, and many things in in the world no longer attract it. Everything they can give it, it has gathered; why should it want to repeat its experience of them? The taste has disappeared because the experience has been obtained; and as this Soul comes back there will be a number of things in the outer world that will no longer attract, and that it will turn aside from with a sense of weariness and disgust. These things will first be the things of the senses, which are the soonest worn out, and it will go more and more towards the things of the mind, more and more towards the things of the intellect, accumulating a larger and larger store for its Devachanic life, a greater and greater accumulation on which it is going to work. So that the life in Devachan will be longer and longer, the Soul working out these greater stores which it carries with it from this earthly life. Coming back then again, having had these long Devachanic interludes, it comes back with this ever increasing distaste for the lower desires, and the links with objects of the sense become feebler and feebler. Its knowledge enables it to recognize the transitory and illusory character of earthly things, and it breaks the links of desire by knowledge; knowing that they pass, it refuses to be attached to them, and so exhausts these links which inevitably draw it back to earth so long as they last. Instead of setting up great numbers of these, it creates only thought-forms of the pure intellect, and the pure reason, and the pure thought, which do not tie it to these transitory things of the earth. And it may break these links in two ways – by knowledge in the way that I described; or also it may break them by catching glimpses of higher and greater realities – the spiritual realities as we say – and that mightier attraction, overbearing the attraction to earth, will draw the desires upwards, purifying them as they ascend.

So that at last all the lower element of desire which is for the lower self will be gotten rid of, and there will be present only the desire to work because the work is useful, to work because the work is duty, to work because others need the service, and so on.

You may thus get rid of the personal element in desire, which is the binding element for return, and in one of two ways; either by a distinctly intellectual recognition of the transitory character of the objects and the exhausting of desire by knowledge, or by the burning up of desire by devotion, and the deliberate sacrifice of everything to the higher ideal of life which is to become its compelling power.

The time will come in this growth of the Ego when it will realize then that the lower earth has nothing which is worth having. By knowledge, by devotion, or by both, it has broken these links. What then will be the nature of the life to be lived, when it is establishing no new links to bring it back to birth? It may be a very active life, employed constantly in working amongst men; for it is not action that binds men to birth, but the desire which causes action. In desire, and not in act, lies this link which draws the Ego back to birth. Suppose then that during a life of very great activity there is no desire; suppose every action that is performed, is performed because it is right to be done; suppose that when it is performed, the Ego concerns itself no more about it; suppose it has no wish for the result of that action, either good, or bad, or indifferent; suppose that when the action is performed there is no link which binds the Soul to it in any way, that it remains absolutely indifferent to the fruit of action, as it is technically called, and works, not because it wants to gain anything, but because it wants to serve and because it recognizes that it is one with the All, and therefore must discharge perfectly everything which the law demands of it in the particular place in the world in which it may be. Freedom of the Soul, then, depends on whether you want to bring about a result by your action, because the result is desirable, or merely because you want to be in harmony with law, because you recognize yourself as part of the All, because you recognize yourself as a channel of the law. If you are nothing more than that, if everything that you do is done because it is duty, if you act neither for pleasure nor pain, neither from love nor hatred, neither from attraction nor repulsion, neither for gain nor loss, then, there being no desire, no link is made: in the doing of the action you are part of the One and the All, and that cannot be bound by these links to rebirth, so that the question of outer activity does not affect in itself the freedom of the Soul. I grant, of course, to the full that people need the stimulus of desire in order to make them act, until they have reached this higher stage where action is perfectly performed for duty's sake. It cannot be reached at a bound, it cannot be reached by its intellectual recognition, it cannot be reached even by saying that it is desirable; it can only be reached by the inner growth of the Ego, which makes it really fundamentally indifferent to all the things which attract the masses of mankind. So long as there is attraction, that is needed for the performance of duty. It is only when the lower nature is entirely the instrument of the higher that a man will lead a life of great activity without the smallest wish to see anything which may flow from his acts; and when that point is reached, he has achieved his freedom, when that is done, Karma for him - save the great Karma of the Universe - is at an end. Individual Karma for him is burnt up, burnt up in these fires of know ledge and of devotion which prevent him establishing any links with the earth, and he therefore makes no fetters which bind him to the wheel of birth and of death. The burning up of Karma in this fire of devotion means that you throw into the fire every action of your life, and like a sacrifice it is burnt up and changed.

Let me give you one illustration only to show you how this change may occur in the higher spiritual life.

There may be a thing which will bring suffering. The Soul which is nearing its liberation is willing to accept that suffering which still it feels; it throws the suffering on to the altar of devotion; the fire of devotion burns up the suffering, and the Soul feels joy in its gift. But that suffering is not lost; it is changed in the fire, and it becomes spiritual energy, which the Great Lodge can use for the helping of man - the voluntary acceptance of pain as a sacrifice to the Masters is changed by that fire of devotion into spiritual energy for the helping of the world. There is the underlying truth of the doctrine of what is called vicarious atonement: not the legal thing that the Churches have sometimes taught, but the sacrifice of a great Soul, which bears suffering and offers it for the spiritual life of the world, so that it shall be changed in the fire of love and come back as spiritual energy to be spread over the whole of the world for the raising and the helping of man.

The Soul, then, thus achieving liberation, comes to the period of choice of which you hear so much. Being free it has a right to choose, and it may either pass onwards into higher types of life, or it may elect to remain within the sphere of earth in order that it may directly help in the freeing of other Souls. That is, of course, the Great Renunciation of which every now and then you catch glimpses in the Theosophical writings; that is the choice of the liberated Soul; it is free, but it remains within the sphere of earth in order to help. It may choose that, by renunciation of its right to go on. It is not bound to earth, but by a voluntary renunciation it remains there with some of the disadvantages, so to speak, which belong to the material existence, for the sake of helping others and carrying on this evolution of the Race.

Where a Soul has thus accomplished its pilgrimage, where stage after stage it has developed mind, where stage after stage it has purified intellect, when it has gotten rid of desire, when it has become a liberated Soul, when it has renounced the going onward for the sake of humanity, when it has remained within this sphere of earth for helping man until the cycle of humanity is completed, then, entering into Nirvâna, there comes the state of All-consciousness, of bliss which no words are able to describe. And then when the time comes for a new manifestation, when the beginning of a new Manvantara approaches, then this Soul which had achieved its liberation comes forth as a Son of Mind, in order in due time to generate mind in a new humanity, to be the Teacher of that humanity in its infancy, its guide in its maturity, rising Manvantara after Manvantara higher and higher. For the pilgrim Soul which began in the germ-union that I described, which went on by accumulating experiences, which then from these experiences extracted their essence, which then got rid of the desires which made it separate, and which unified itself once more, becoming a unit consciousness in a mysterious way which can not even be sensed until at least the lower grades of the higher consciousness have been experienced during earth-life by rising out of the body and learning what it is to be an Intelligence working without the shackles of the brain - such a Soul thus having worked through its pilgrimage and regained unity shakes off the compound individuality, retaining the essence of it which it extracts; being a unit it is incapable of disintegration, it is for ever immortal – the Soul has achieved its immortality, and through all Universes to come it is one of the Workers, one of the Builders, one with God in work for the worlds.

What is Theosophy?

With Special Reference to Scientific Thought.

An Address given in London on September 4th, 1891, to the United Democratic Club.

IN the lecture that I propose to deliver here to-night, I shall try to put before you as plainly and as clearly as possible the leading ideas that go under the name of Theosophy. I shall not, bearing in mind the audience that I am addressing, try to win you by mere skill of tongue, because I know there are too many here connected with the Press to make it worth while to try to get round them by words which do not convey any meaning. I shall make the lecture as terse as I can, for I want to get a good deal into the space of time that I have at my disposal, and I shall endeavour to put those ideas in at least a coherent fashion, which will lay them open to attack and discussion on the part of those who do not agree with them.

Theosophy is an extremely old theory of the

Universe and of Man. It starts with the idea that the Universe and Man are primarily spiritual existences; that what we call spirit and matter are not really two and in antagonism, but are one substance in different stages of evolution. If you go back to the beginning of our present Universe you will find as you go backward that life becomes more and more spiritual in its manifestations, that is to say, the denser forms of manifestation are later in time than the more subtle and ethereal forms. But the principle of the Universe is Life, and not lifeless matter or energy; primarily life and consciousness are at the core of the Universe, but this body issues forth in various forms of matter, the forms becoming denser and grosser as evolution proceeds, and that finally when this first stage, the plane so to speak of the Universe, is complete, the whole is ready to start along the plane of further evolution. You have a Universe manifested in seven different forms of matter, subtle at the beginning, denser at the end, and, so to speak, arranged in seven different stages or, as we say, seven different planes of manifestation. To take a very rough image, to put to you by physical analogy what I mean, you might have in a chemical experiment a receiver that appeared to be empty, you might subject that receiver to constantly increasing cold. As you lowered the temperature the apparently empty receiver would soon be filled with a delicate mist; as you continued to lower the temperature, that mist would gradually assume the form of definite vapour, then of definite liquid, and further on of solid. You would have but the one substance right through, but you would have had it manifested at different degrees of density according to the conditions under which it was manifested, and so we say that the Universe is but of one substance essentially, but is manifested in different forms according to the conditions in which this substance has been bodied forth. I am simply putting to you the statement without for a moment giving the arguments with which it is supported. I will ask you simply to have clearly before your mind this conception of the Universe in seven different stages or planes of various degrees of density, so far as what we call matter is concerned. Each of these planes has life manifestations suitable to the plane on which they are manifested: whether or not the organisms of one plane are conscious of the organisms of the other will depend on the power of sensation possessed by those organisms. If you take Man as you have him now he comes by his physical sense into contact with the physical plane of the Universe. He is able to see because the molecules of the eye can vibrate in unison with those waves of ether that impinge on the mechanism of the eye, he is able to hear because the organ of hearing is so disposed that it vibrates in unison with the waves of air. He becomes sensible of the material universe without him because his body is able to vibrate in unison or in response to the various physical vibrations of the physical universe with which it is surrounded.

Let me now take the next stage. Man, like the Universe, consists of seven principles in seven different stages of this manifesting spirit-matter or substance. We say that these seven principles existent in man are correlated to the corresponding plane of existence in the Universe, and that just because he has in himself these seven different stages of existence he is able to investigate the whole Universe around him, becoming conscious of each plane in the Universe by virtue of the corresponding principle in himself. So that in the theosophical conception of Man and the Universe you have two images, so to speak, that respond the one to the other, as the image of Man's physical body might be reflected in the mirror before which he stands.

He can know the Universe because he is himself the Universe in miniature, and as he develops in himself each different principle of his nature he is able to investigate the plane of the Universe to which that principle in himself responds. Now, it is by virtue of this fact of man's nature that knowledge becomes possible of these different planes. Take for a moment again the body: you can investigate the physical plane of existence by your physical body, but beyond that physical plane you cannot go with your bodily senses. Now, for a moment, as an hypothesis, suppose that there is a subtler form of matter than the matter that composes your bodies, it is not at least an impossible hypothesis, science is always discovering rarer and rarer forms of energy, and the latest discovery is not susceptible to the ordinary senses. There are light waves which are too rapid for the molecules of the eye to vibrate in response to when they strike upon them. The eye is, therefore, unconscious of their existence. It is idle to say that these light waves do not exist; their existence has been proved by scientific men, partly by observing them as they affect other forms of life and chemical combinations, and partly by subjecting them to experiments which, by changing the rate of vibration, render them susceptible to our senses; therefore, they do exist, as a matter of fact. Now, the Theosophist says that the next set of vibrations, too subtle and too rapid to make any impression on the physical body of man, are on the Astral Plane. It is possible to develop in Man a form of nervous sensitiveness which will enable him to respond to those vibrations, just as a physical body responds to the air. You can throw Man into a special nervous condition which will render him far more sensitive than he is normally; when he is thrown into that condition you will have evolved a sensitiveness that responds to those more rapid and ethereal vibrations, and then those vibrations become as real to him as light is to ourselves. That condition is the condition variously known as hypnotic trance, or mesmeric trance, or condition of conscious clairvoyance. In America, where the climatic conditions are different, people are evolving normally this increased sensitiveness to peculiar external conditions. In England, what is called the "psychic" is comparatively rare; they are not unknown, but they are regarded as "cranks." In America, where the air is brighter and drier, where the conditions of life are more rapid and the nervous system has become more tense, you have the psychic development carried on to a much greater extent and developed among a much larger number of people than in Europe at the present time. When we want to obtain these conditions here, we mostly do it by shutting the ordinary senses and by making the body impervious to the impressions of the physical universe outside. You render the person blind, deaf, and insensible to touch from without. When a person is thrown into the mesmeric trance, you may beat a gong in his ear without his hearing it, you may flash the electric light into his eyes without his seeing it, you may run needles into him or subject him to shocks from an electric battery, and the most delicate apparatus of the investigator will not indicate any response to these conditions of stimulus from without. I am putting a matter of knowledge ; this is a fact testified to over and over again by every scientific man who has investigated hypnotic phenomena.

When the body is in that condition, closed to every ordinary stimulus from without, you obtain a completely new set of results – the patient is more keenly alive to the person who has hypnotised him than he is normally alive to the persons who surround him in ordinary life. That one person can communicate with him and produce extraordinary results. He can make him describe objects that have no existence outside the thought power of the hypnotiser. He will place an imaginary object in his hands, so that the man cannot put his hands together, thinking they are stopped by the object he is holding ; he will describe elaborate pictures on a blank sheet of paper, those pictures having no sort of existence so far as the ordinary eyes of the unhypnotised person can discover. These facts point us to a whole plane of existence of which we are normally unconscious, and the theosophist will tell you you have transferred the consciousness of this person on to the Astral Plane. where you are dealing with matter invisible to the normal senses and with vibrations of matter too rapid for the ordinary senses to perceive or respond to. When you have passed into this Astral condition, you are able to see with fresh sense of sight and to hear with fresh organs of hearing, and these organs of sight and of hearing function under laws which are very different from the ordinary law under which matter works in its grosser manifestations. You can see to a distance which otherwise would be impossible, and you can see through objects impervious to the ordinary eye. You can take a board an inch and a half thick, you can place this in front of the eyes of the hypnotised person, and on the far side of the board you can place ribbons of different colours, the hypnotised person will see the ribbons and tell you the colour of each ribbon. Under these conditions, the sight that you have evolved in your patient is a sight which is not trammelled by the ordinary material conditions under which you are accustomed to let your organ of sight function. There is nothing in that particularly remarkable, because, if you are dealing with vibrations so subtle that they can pass through interstices of gross matter, then you have the conditions of eyesight present under the responsiveness of more delicate sense, so that you are still within at least the analogy of science when you are dealing with this second plane. When you go a step further you come to the plane of the emotions and the feelings, and you can transfer man's consciousness to that plane of existence, so as to render him unconscious of an injury received to the body. Take the case of a soldier wounded in battle. Any soldier thus wounded will tell you that it is not until the rush of the fight is over that he becomes conscious of bodily pain, but, if the consciousness were then in his physical body, he would know then the pain ; but consciousness has passed from the physical plane to the plane of emotion, and not till the passion is stilled will he recognise the pain.

Beyond this, there is the plane of the mind. The plane of the mind is still a plane of matter, for I will ask you to remember that there is no essential difference between mind and matter; they are one substance manifesting under different conditions or planes. Mind is still far subtler, far more ethereal, but it is as real, and we say that the mind functions on that plane; that, when you think, your intelligence is a force or an energy that is working on this mental plane and that its workings there are as perceptible to the mental organism as any workings on the physical plane are perceptible to material eyesight; that, when you think, you create a thought image and that that image may be rendered visible under certain conditions of enormously increased sensitiveness and that it is possible and has been done to so develop the mind element in Man that you can separate it from the brain organism through which it normally functions, and free it from its restraint. You can make it more active and more potent, just as your limbs are more active if you take off them a weight of chains. People can develop the power of transmitting thought through this thought-medium from intelligence to intelligence without the ordinary material mechanism that you normally employ for that purpose.

But I am still only dealing with things that Science is beginning to dream about. Twenty years ago, to talk about conveying ideas without the use of written messages would have been to render yourself a candidate for the nearest lunatic asylum ; but only a few weeks ago, before that eminently respectable body the British Association, it was confessed that here there was room for investigation, showing that such communication was likely to be possible, and Professor Lodge even said that it might be taken as almost having been proved possible. It has been proved over and over again, and, just as your scientific men scoffed at Mesmer and fifty years later invented the new name of Hypnotism, so your science to-day is beginning to recognise the necessity of investigation into this new means of communication and this working of the intelligence in subtler realms than hitherto it has admitted.

All the fuss made during the last week about the possibility of communication in various fashions is nothing more than a mere outcry, but people imagine they have got hold of a miracle. If you installed an electric wire across the Desert of Sahara, you might very much astonish some of the natives by communicating across the desert in a way that they would not understand, or you might even flash a message by utilising the sunbeam as the method. It is only going a few steps farther to be able to control other forces without your physical apparatus, and even without the sunbeam to help you. How people would have laughed a century ago to hear scientific men say they would be able to converse by means of a wire! It is further, I admit, along the analogy of scientific thought when you are able to deal with subtler currents, but no more miraculous than any other form of dealing with natural forces, and it is only ignorance that cries out "miracle" or "fraud" – miracle where there is the superstitious to account for the unknown – fraud where there is the gross ignorance that denies the possibility of further advance, and because the nineteenth century is so wise and thinks that none can be wiser.

Let me here allude to one phrase which will show you at least in the thought of those who believe in the existence and the control of those forces how thoroughly natural they are. Some years ago in India, when H. P. Blavatsky was there, she was utilising a number of those forces in connection with her Teachers. On one occasion she was asked to bring about a certain result. I forget whether it was the conveyance of a letter or some other object. She was asked to convey that object to a particular spot within a particular cushion, and she said she would try to do it. Later on in the same day, talking over this matter, she was asked, "Will you change the place that we have already asked you to deal with and put it in another place?" She asked, and the answer came, "We have set the currents to the place where we were asked to set them, and it is easier therefore to send the object to that spot." You set those forces going and, unless you are a miracle-worker, which the Theosophist does not believe in, you must work under natural conditions. If you have started your forces along a line, that is the line along which you must follow; if you want to alter your result, you must alter the cause by which your result is to be brought about. When you begin to learn anything about these forces, you find you are in a world of law as strict as that of a chemist; you can no more produce a result which is not within the law than your electrician can produce an effective spark from the machine if the atmosphere is charged with moisture. It is this side of Theosophy which tends to prove most attractive to those who are simply curious for a new sensation. They don't care twopence about the philosophy, they care even less about the Theosophy, they want something that will sell their papers ; something they don't understand is the one thing that they want. In the *Daily Chronicle* all that I told them about the Mahatmas was put down.* They carefully cut out what I said about the results ethically of the teaching of these same Mahatmas. (Shame !) I don't say it is a shame ; I am very glad to get half of it in ; the great difficulty that every new theory has is to get a hearing at all ; when they came to the end of the column, there was a battle or a murder or something, so my other half had to be omitted. As far as the *Chronicle* is concerned, it has acted in a most fair and just fashion and has given a fair hearing to both sides. I am not putting that in any spirit of complaint, but as showing you one of the disadvantages under which really serious people labour when they are dealing with a philosophy that has a side that appeals to curiosity and the desire for experiment and what is known as phenomenal, but which has to us a far more

* The reference is to a discussion on Letters from the Mahatmas in the *Daily Chronicle*, in which Mrs. Besant took a leading part. —*Ed.*

serious side in its bearing on the progress of the race.

Let me show you how this view of Man and the view of the Universe passes onward into our philosophy. These discoveries about mind – the power of mind to make thought images, to transfer by the subtle forces under its control ideas from place to place – we say that these show that every man has in him the same power of creation as that universal mind of which we say the Universe at large is the manifestation. I don't mean by creation making something out of nothing – I mean by creation utilising the various forms of matter around you to produce a new result, just as you say the artist creates a statue, just as you say the musician creates the harmony when he by the efforts of his genius gives some great opera to the world, so I speak of creation when by the subtle forces of the mind you utilise the forces and the material around you to bring about certain results that you have made up your mind to effect.

Man is essentially a thinker, and that thinker is imperishable and eternal. We say that the thinker is the real man, and that the body is the instrument the thinker uses : a more or less good instrument as may happen, so that the brain, through which the intelligence or the thinker normally functions, may be a better or worse instrument for the expression of that thinker's thoughts, but always an instrument by which he must work on the material plane, and only as he is able to transcend that plane can he work in the subtler media of which I have been speaking.

We say that the speaker passes from incarnation to incarnation, moulding and building as he goes his own future and the future of the world. We say that in every thought that you think you create a thought image on the mental plane and that the whole of your life is a constant action of creation of these thought images, that these images persist, and that during life you are constantly adding to them, that they act and re-act on each other like any other forces, and that they build up the human character ; that your character is made by your thinking, and the *modus operandi* of making it is that every thought creates a form on the plane of thought, and that the aggregate of these forms working upon each other makes the character of the man at the end of his life-experience. We say that that mental thought image, that is the outcome of the life thinking, persists after the death of the body, that it does not depend on the form of matter you call physical, it depends on the subtler form of matter I have been speaking of, and of which Science is beginning to get evidence in thought transference. When the time comes for the thinking principle or the thinker to again clothe itself in material body, this thought image, which is the outcome of the previous experience, is the mould or the matrix in which that physical body will be built, so that what you call the character with which a man is born, the tendencies that every child has when it comes into the world, the tendencies to act one way rather than the other, to be quick or to be slow, that all these tendencies that you speak of as inborn character are the natural and inevitable results of the thinkings of past incarnations which have made a mould of subtle matter into which the grosser matter of the material body is builded.

If this is true, then its bearing on life and conduct is enormous, because it means that everyone of you is creating at the present time his own future and the future of the world. If you are thinking selfish and evil thoughts you are making a form of matter that outlives your physical body, that keeps the form that you imprint upon it and into which the physical molecules of your future life will indisputably be moulded so that those who think selfish thoughts habitually will in their next incarnation be born with a selfish character, so that what a man is now is the record of what he has been in his past, and the true life of the man is in the thought and not in the act. Whether a man is a thief or a murderer depends on opportunity. It is not every thief that gets put into jail. The thief is the man whose tendencies are thievish and who tries to get that which he has no right to; whether that thievish tendency works out in act depends upon the opportunity, if he has the chance he will be a thief in act; but his moral value depends on the thought. He is judged by the thought and not by the act. Many a thief in Holloway Jail is not as deep-dyed a robber as the man who poses through life as a splendid character. This bearing on our theory of right is to us the most important part; that is what we call Theosophy: the great central truth of the Universe and Man. And this doctrine of Reincarnation is not only the basis of our ethics as regards the individual man, but the foundation of our belief in Universal Brotherhood: it implies the essential equality of Man.

The differences in material condition are mere outward accidents and do not touch the inner self, whether a man be prince or pauper, whether sage or sinner, the essential unity of humanity remains, and makes them brothers in fact, whatever the dissimilitude of the outer garb, and this is the basis of human brotherhood. What matters it to you and to me that in this particular life there may be an apparent difference between us ? We are one in that we are human, and the intelligent thinker is the same for all, although in different stages of evolution, and it is quite possible that the man whom you despise as stupid, as profligate, as vile, may be a human being further in evolution than yourself ; although for the moment he may be under passing disadvantages. The progress may be clogged for a moment, although he may be advanced in the whole evolutionary cycle. The passing nature through which he has to work may be some gross tendencies inherited from a past that he has not conquered. It might well be that such a person is only under temporary obscuration, and is really a far nobler type than the man who judges and despises him, and who may not have travelled as far along the road as the one for whom he feels contempt. Theosophy teaches you to be careful in your judgment of your fellows : you may say a person is repulsive, covered over with some horrid skin disease, better so than have the disease driven in, so that he becomes a source of infection; so many a man in the outer vices may be working off the symptoms of evil in him, and may come out the nobler. I say that because sometimes you find acts of the utmost heroism in those who have seemed even the most degraded. We tell you that no man is wholly evil, that at the root of things man is good, and not evil, that he will work through the evil to the light; though he is bad to-day, do not make his road rougher by putting your hatred to keep him down, give him the hand of help to lift him upward and so live through the vice which makes him hateful to the many to-day; and that is the ethical side of our teaching, but it depends upon the philosophy. You cannot work it out in that fashion unless you accept the central doctrine of re-incarnation, and you have no reason to accept that unless you can work it from plane to plane. When you can prove intelligence working apart from your material organism you have taken the great step which makes the whole of our theory intelligible.

We, who are Theosophists, do not encourage people to rush hastily into a mass of experiments in which they are very likely to lose their heads and destroy their nerves. I am not what is normally called a Spiritualist, because I think that Spiritualists make a mistake in the deductions from their phenomena. A large number of the results they get are genuine, though there is a large amount of fraud as well. But, unless you study the subject with knowledge, you are likely to ruin your brain and your nerves. The attitude of reception to all outside influences which is necessary for mediumship - because you must render yourself specially susceptible – that study persisted in month after month brings about an abnormal condition which is likely to ruin the health of the person who subjects himself to it. And so with hypnotic experiments : you may very easily destroy both nervous system and moral character by playing with forces whose management you do not understand. Now, what is necessary before you begin to experiment is study : study the theory before you practise. You would not let a young lad loose in a laboratory to put together whatever compounds he might choose ; you would say, "You must not try these compounds until your knowledge of them and the laws under which they work is such that you can make your experiments without laying the laboratory and the houses all around in fragments." When a person comes and says, "I want to know this experiment," we say, "No; if you want to know you won't mind the trouble of studying." The curiosity which runs out to see the Queen, or a jumping flea, or a fat woman, is not the kind we want for this work. We quite admit that you can get knowledge of many of the occult arts without moral character, you can become a mesmerist without the slightest attempt to control your passions, but we say that those of us who are studying the matter seriously won't help you to do it. Find out what you can, but don't ask us - who have pledged ourselves to serve the race before everything else, who have pledged ourselves to utter subjugation of self before we place our hands on one of these forces, who have proved the truth of the pledge by living lives of self-denying labour for the people for years before we tried to study these things at all – don't ask us to take you into the occult laboratory and enable you to experiment with the most explosive substances before we know you will use them for service and not for self.

Take the ordinary man: he loves his wife and his children more than the children in the gutter, but he must not be an occultist while he has one love for a human being which will make him unjust to anyone else. As long as he would rather serve his own child than the child in the gutter, so long he has no right to use these occult powers : he might use them to serve his own child to the destruction of others. The claim on you is the claim of want, and not the claim of personal affection ; and if the gutter child is starving, that child has first claim upon you because his need is greater. I don't say that this is the view that the mass of the people should take – it is not. But you cannot enter into the occult school until your love for the race has become as pure, as passionate and as self-denying as your present love for wife or child, and until that be so, take the evolution of the race as it goes, but do not claim to climb the mountain up which only those can climb who have thrown every personal desire aside. All the race will come to these powers in due time ; every son of man will be born into this heritage of absolute royalty over Nature, but he has got to evolve into it, and if he wants to evolve much faster than the race, he must pay the price ; and the price is, to live the impersonal life, so that he may be a safe custodian of those tremendous powers of Nature. That is the school through which everyone has to go, and I know no reason why a curious person should escape it more than anyone else. It won't do him any good to escape it, for if he happens to be a person of psychic development, you will have given him the clue to get mastery of that power and he can use it, not for a harmless purpose, but for the mischievous purpose of slaying an enemy. Already you are beginning to learn something of the doctrines of hypnotism, that is the very lowest of those powers possessed by the occultist; yet hypnotism has been used for criminal suggestion, and the true criminal has escaped, because there is no law which can touch him. Is it altogether so foolish then, this secrecy?

People say they will not believe me. I don't mind whether you believe me or not. A fact of Nature does not alter whether you believe it or not. My only reason for ever mentioning the letters from Mahatmas in public was, not to show that letters could be precipitated, but to show that my friend was not the forger she was stated to be. I have had letters from the same person, and it proves that she did not write them. If the writing was the same, it was not the hand of H. P. Blavatsky, therefore she did not forge that writing. It was a perfectly fair point to make in defence of a woman who has made my life all that is worth living.

I have had letter after letter from good

people, saying, "You are deserting the poor." I deserting the poor? I am learning to serve them better than I ever served them before. I have given up the momentary praise which comes from rescuing one woman here and there, a solitary unit out of thousands who perish, and I am learning how to save the thousands. I am learning how to use my brain so as to make it more serviceable to the wretched than it has ever been before. I am giving up that political work that only deals with facts and not with the causes by which they are produced. I am doing so, in order that by a more complete self-devotion I may rescue those who are perishing before my eyes at the present time. Hundreds can do my work on the School Board much better than I can: let them do this work that they can do as well as I; let me, who have no human tie to keep me back, except the tie to my race—every other tie having been broken by the force of circumstances—leave me free to go along the road where few will care to follow me. To learn the lessons that can only be learnt by the surrender of everything that makes life precious to the many. Believe me that in giving my life to this new work, I do it, not for selfishness, but for self-surrender, and in the hope that in long years to come I may realise the idea that I formed in childhood, and that I have nurtured and tried to work out in womanhood—to save the race of whom I am a part, to learn lessons which worked out in life will render the present social misery impossible for evermore and make the world worth living in, to help the miserable to whom my heart is given, to gain a ransom for my brothers and not a selfish bid for mere knowledge or self-advancement.

The Evolution of Man.

An Address at the Parliament of Religions, Chicago, 1893.

IN finishing this brief description of some of the leading Theosophical teachings, I have been desired to take up and deal with the Evolution of Man. Man, as you take him in the past, man as we see him in the present, man as we shall see him in the future, the very first fruits of that future being men living on the earth to-day.

Evolution in our modern civilization has been a word of power over the minds of men. Those glimpses that the West has got of evolution give us but half the story, draw for us but half its circle, and with a half truth give us an unintelligible, inexplicable mystery of a life that comes from no centre, that finds no intelligible goal. For just as we see in our Western evolution that life appears, that a certain interaction of force and of matter has out of death made life, that out of unintelligence springs existence.

that out of the brute springs the man, so evolution springing from the lower stages of life is to pass onward and onward to an end as emotionally, as intellectually unsatisfactory as its beginning is vague. For in the latest presentments of science we are told that in this chain of evolution the latest link shall be as low as the first ; gradual retroaction, gradual degradation, until worlds evolved only from matter by energy shall resolve back again into uninhabited desolation, either burned by fire or frozen into obliviousness of life, until, disintegrated once more, they will be built up again in the far-off future of existence.

Such an evolution, were it true, would be the dreariest theory of life that human mind could conceive – unintelligible to the brain, unsatisfactory to the heart. Far other is Evolution as we have learned it from the ancient books, as it has been traced for us by the Masters of Wisdom ; for they tell us of that spirit, to a description of which we have just been listening, out of which springs a universe, the universe passing back, full of life, to expand into the Divine All-Consciousness. They tell us of an Involution which is the Source and the Fount of Life. Spirit involving itself in matter that it may become the mainspring of Evolution, and may gradually mould matter into a perfect expression of itself. And then this descent of Spirit into Matter - this expansion of Life from within, passing through stage after stage of evolution, reaches its lowest point in the Mineral Kingdom, thence begins the long climb upwards, thence, by expanding energy, we can pass onward, stage by stage, to the early evolution of Man, Man as he appeared in the present phase of the earth's existence, first of all living things, the pattern of all forms, containing every possibility that that stage of the evolving globe was to produce. Passing from stage to stage, till the animal body was builded, till the astral form into which the physical was moulded was ready to gather the physical together and make a possibility of material human life. Then in that focussed the life energy of the world, gathering to itself the forces which knit the molecules together and co-ordinated all into the astral and physical bodies. And then as the last touch of animal man, of this lower and transitory existence which was to be the garment of the soul, we find appearing the passional, the emotional, the instinctual nature, that which Man has in common with the brute, and out of which in course of evolution that part of the brute nature also took its rise. So that we come to a stage of human evolution where the animal side of Man is completely builded; the tabernacle of the flesh is ready for its tenant; the house of the soul ready for the incoming mind, and Man at this stage of his existence nothing more than a beautiful animal should appear in the possibilities of adaptation built into a similitude that would be able gradually to be moulded by the indwelling soul into a perfect instrument for expression on the lower plane of life; and then, to that abode builded for the mind, comes the thinking entity, that is, the real Man – Man whose very name comes from the root that means thought, Man whose very name in our own tongue is identical with the Sanscrit word which is the root of thinking; so that in our very title in the world we bear the impress of our special characteristics, that the human soul is the thinking energy. The thinker that makes the complete Man a possibility came not from the lower world, not given by material nature, not evolved from the astral plane, not given birth to by the lower life, not taking its origin in the passional, the emotional, the instinctual nature – Man's soul comes from above, not from below, not climbing upwards from the brute, but the focalized reflection of the Spirit.

That is the soul that came to Man as animal

and took him into his charge, to build him up to the divine: for this thinker is the God in every man, the God who has evolved from Matter, the God who has descended that he may subdue to himself the lower nature and render every plane of existence translucent to a vehicle of the Divine. This God in Man is the teacher, the Guide, the instructor, the Helper, and also in his lower aspect the gatherer of experience out of which he shall build up character which he shall carry back with him to the higher work that lies before him in periods of existence yet unborn in the universe, that are still in the obscurity of eternity.

This thinker, this God descended into matter, has a dual aspect, one face turned to the Divine which is its source, the other face turned to matter which he has come to dominate and to subdue. These are the higher and the lower minds, the rational, and, in its union with the lower nature, the animal soul in Man: so that in its double nature you have the aspect that is turned to the brute to train it; you have the aspect turned to the spirit that strives ever upward towards union with the purely divine. And the whole life of man is the battle-field of that dual nature – the God struggling with the brute, in order that the brute itself may become divine. That is the way that Man evolves, that is the building up of the divine in the midst of the earth on which we live. Do you doubt that God is in every Man? Do you doubt that the essence of humanity is divinity itself? Men talk of others as sunk in evil. Men speak of their own race as corrupt, and by the very degradation they ascribe to it they make it more degraded than otherwise it would be; for we tend to reproduce the opinion that surrounds us. If we are evil and brutal, we tend then to take on, as it were, the character which is ascribed to us too often even by the religious faiths. But if Man be divine, if the very heart of Man be light, then you can appeal to the divine within the lowest, and know that answer will come, however muffled be the veil of flesh. Would you have proof that God in Man is present in the vilest, present in the most degraded, present in every son of man whose life seems that of the brute rather than of God? Come with me to one of our English villages far away from the ordinary haunts of men – a village which, once all beauty, has been defaced by the greed of those who possess it and the carelessness of those who live in it. We have some mining villages in our country, I am ashamed to say, where the lives that are lived are lives of the lowest, of the most ignorant and most degraded. Not all of our mining population are thus. Some of them are strong and self-reliant men, but it is not of them I am thinking now. I am thinking of some villages I know where if you walk down the village street you would find gathered in front of the public-house men whose language soils those ears that hear it, who speak foul words, who are gambling, betting, drinking, finding all pleasure in the senses, and you would say, "No light of the divine is there." Are you so sure? Wait and watch them as you wait, and as you are there, thinking how degraded men can be, how they seem to be nothing but the vilest of living creatures - listen to a sound there that makes every man spring to his feet, in order that with every sense alert he may hear the sound distinctly and understand what it means. There is a far-off rumbling that seems to shake the ground on which they stand. The far-off rumble that comes louder, louder, louder, till with a mighty clap as of a thunderbolt there is a crash, a roar, and a pillar of smoke that comes up from the earth, and from mouth to mouth the word flies, "Explosion in the mine below," and men are there, living or dead, one cannot tell. In a moment the whole village is alive, men, women, and children rushing to the mouth of the pit. There are cries of women who know not if they are wives or widows, wailings of children who know not if they be fatherless, and the strong men gather around the pit, the pit that is black with smoke, and unheeding that fiery death that is beneath – there is a struggling at the mouth of the pit, men struggling with men, and struggling for what? Come near them and you will hear the words that flow from their lips. "Go back, you've got a wife or mother. Let me go down who have none to care if I die." And the men who were swearing, who were gambling, who were drinking, hearing that cry of men in agony, forget their brutehood and remember the God that is within, and they fight to go into the cage, they struggle for the chance to sacrifice their lives for their comrades; and down they go, down into the hell of the burning mine, to see if some comrade be there still with the life within him and they can bring back to woman or to child the bread-winner of the family, the support and guardian of the home. Do you dare to tell me those men are not divine? Do you dare to say that where sacrifice is pleasant, the very source of sacrifice is absent from the heart of man? I tell you there is none however degraded, none however ignorant, none however vile, in whom the divine spirit has not His Sanctuary in the innermost heart, who shall not at length become pure as the little child with love that raises him from the mire of sin, and that energy of divine life which has in it the promise of triumph, however far off that day of triumph is. And Man evolving by this inner force, life after life, makes slow progression till a time comes in the life of the man when more rapid growth begins to be possible ; the time when the man by gradual evolution is beginning to understand the far-off possibility of reuniting, as it were, the higher and lower mind. When the upward striving of the lower mind is beginning to reach by aspiration that higher one of which it is the ray. When the higher mind, having worked for ages in human evolution, is beginning to be able to impress itself on the tabernacle so that that tabernacle is conscious of the indwelling of the God, and then there comes a time when the man thus evolving begins consciously to set before himself a definite aim to bend all his efforts in a definite direction, and there will be evil in the heart only or in the heart and lips as well, yet a conscious acceptation of Man's true goal in life, the service of his race and the giving of himself for Man. And then the man who has reached that point in evolution vows himself to the service of all that lives, and puts before him for all future lives that may come to him the one object of growing so that he may help others, of learning in order that he may enlighten their ignorance, of strengthening himself that his strength may be of help in raising the world of which he is a part, and then the lives consciously directed become more rapid in their evolving energy, life after life adds more and more rapidly to the vision of the soul, to the power of the lower mind to respond, till, stage by stage, the story grows deeper and higher, till step by step the life becomes purer and purer and fuller; and the last cycle of births is entered, which when completed will leave the man one of those who have triumphed over sin and death; and when these last lives are beginning, one lesson comes from those who have already achieved, one special direction is given to the disciple by which his life is to be guided, by which his safety on the path is to be secured. You may read it in the same book that I quoted this afternoon. Those fragments of the Book of the Golden Precepts, that are the very hymn book of every true disciple, and there you will find that the law of life must be compassion, that the law of life must be feeling and suffering and enjoying with others, that no tear must be allowed to fall till the effort has been made to wipe it from the sufferer's eye; but that every tear not wiped from the eye of a brother must remain burning on your own heart till that which caused it is removed. An then these lives of continual effort for others bring at last the evolved Man to the point where perfection is reached and triumph over death secured. They lead him to the point at which, once more to quote the same book, "He holdeth Life and Death in his strong hand." He is no longer a disciple, he stands complete in knowledge; he is no longer a combatant, the victory lies behind him and the spoils of victory are in his grasp. What shall he do with them? How shall he spend them? Weary with ages of struggle, what shall be his final choice? He stands on the threshold of that world, separated from ours by difference of condition, which no bridge is able to span. He stands on the threshold of that state of consciousness, so misunderstood in the West, called Nirvana, that mighty state of all consciousness and all knowledge which no words can syllable and no heart of Man conceive. He opens the door that leads to that sublime condition. It is his by right of struggle; it is his by right of conquest. His very foot is on the threshold of the doorway, and one moment he pauses ere he crosses the threshold. And as he stands there, Lo, a Voice, the Voice of Compassion itself, sounds in his hearing, and he pauses to listen. "Shalt thou escape while all that lives must suffer? Shalt thou be safe and hear the whole world cry?" And in the silence that follows, the cry of the world is heard. Across the abyss comes the sob of humanity, orphaned humanity, that is without guide and helper, and that sees one of its greatest passing out of sight. All the cries of men in agony, all the shrieks of women trampled under foot, all the wailings of little children in our world, make one mighty chord of anguish, and they cry to him to stop. What has his life been for many a life past? It has been a life hearkening to every cry of pain that comes to it. It has been a life that responds to every appeal for help that reaches its hearing. All the life has become divine compassion. Can it be deaf when help is needed by men? And in that silence, broken only by the sob of anguish, in that silence is made the great renunciation, the door of Nirvâna is closed by the hand that opened it, from the threshold that might have been crossed the foot is withdrawn, and the Master turns back. He chooses the great renunciation, he chooses voluntarily to live in the world for the helping and the guidance of men. He brings back the strength he has conquered, the wisdom he has gained, the love that is his very nature, and he lays them all at the feet of humanity that he is willing to serve – his knowledge for its guidance, his purity for its cleansing, his strength for its uplifting, his infinite compassion to have patience with its folly, forgiveness for its wickedness, endless endurance till it learn wisdom also by experience through which it passes. Those are the men that we call Masters. Those, the mighty souls to whom we give our heart's homage, not because they are wise so much as because they are loving; not so much because they are strong as because they are Compassion absolute. Those are the guides and the teachers, those the examples that stimulate us to work.

Behind the movement which we have been considering for the last two days stand those servants of men, inspiring all that is best and noblest in it. I do not mean guiding its policy, I do not mean driving it along every step of its life, for they let their servants learn by their own mistakes, desiring not mere puppets that they control, but men and women evolving toward perfection. That is the strength that lies behind our movement as behind every other great movement for the spiritual good of men, for it matters not whether we know the Masters – they know us. And they give their help to every one who works for Man, no matter whether his eyes be blinded or whether they be opened to the light they shed. That is the secret of our strength. What is it that in this Parliament of Religions has drawn crowd after crowd at all its sessions, to learn the truths that a few amongst us have here been employed in imperfectly setting forth? Youngest, you may say, of any movement as the world knows us, though in reality the oldest of all, what is it in this Theosophical Society, not yet in its twentieth year of life, which is making the eyes of all men turn towards it and making the hearts of all men ask what it has to give? Men are hungry for spiritual truth. Men are longing for spiritual knowledge. They ask for a knowledge of the soul which shall not be based only on faith; for a guidance in life clear and definite that may satisfy the heart and the reason alike. And this movement was started by those Masters of Wisdom to feed the hunger of the soul which the cycle of time had brought round again; and they sent into the world their messengers that they might make this movement possible. Who is its true founder so far as the material world is concerned? They selected, these great souls that stand beside this Society, a Russian woman, outcast from home and friends, Helena Petrovna Blavatsky, who went out from her Russian home, leaving wealth, rank, princely position behind her, eager only for knowledge of the truth and union with the divine life. Through many a land she travelled, through many a clime she wandered, one after another she examined the teachings of the world, till the eye of the soul was opened and the Master she served sent her out to do his work. Penniless she came back to the world. Told to go to America, she went to France – as far as her money took her – and there coming into possession of a few pounds, enough to land her in New York, but no more; yet nothing could stay her. She went with the word behind her; that word he gave her; and she came alone and friendless to your country to face the materialism of the West and to proclaim as alive again the true and ancient Wisdom-Religion. She was scoffed at and derided, laughed at and defamed. Every foul word that the malice of foul minds could image was heaped on her one head. They never thought she was a woman and had none to help her, and they did right, for that lion-heart asked no sort of consideration, and she would not use sex as defence against cruelty. She lived her life, she gathered round her men and women who got some glimpse of the strength that was within her, and the beauty of the divine life that she enshrined. They tried to crush her with calumny, tried to destroy her influence. What is the answer? The answer is that two years and a-half after she passed away there are thousands of men and women scattered the world over to thank her for the life she lived, for the guidance that she gave to life. They thought they had crushed her with their Hodgson babble; they thought that they could crush her with all their Psychical Research Society Reports, and the answer is that we are living to-day and we stand as testimony to her work, as witnesses to the life she has made possible for us. How has such a movement spread? How has such a Society been possible? Because of a spiritual life that lies behind it that no slander can wound and no power of man can touch. And to-day, to-day, those who made the movement possible glance over the Western world to see where some souls may be found willing to be helpers with them in the redemption of humanity, willing to share with them in the toil and triumph that lie behind. Here and there

there is some soul that catches glimpses of the light that shines from behind the veil, and gives itself in its pure measure as they had given themselves for men. Such are the helpers of the Masters. Such the co-workers that they are seeking, and not one of you but, if you chose to take the higher path, might make to-day your first step along the road, a step, it may be, feeble, uncertain, and halting, but if made out of love to Man and devotion to the spirit has in it the certainty of final success - is the beginning of the journey that shall lead you to be co-worker in the spirit. That, then, is the final appeal that from this platform comes to every man or woman ready to give himself for the helping and the saving of man. There are so many that want help, is there none to give it ? so few to speak for the spiritual life among so many that are sunk in the flesh. And this I say to you, that no joy of earth, no hope that gilds an earthly future, and no delight that comes of earthly triumph; no such joy, no such happiness, no such ecstasy, bears any more proportion to the joy of the spiritual life than the fog that surrounds some mining village is radiant as the sunshine, or the pettiest joy of the gnat in the sunshine can emulate the power and delight of the intellect in man.

For greater than intellect is spirit, brighter than Mind is the Supreme Life; one joy, one peace inexhaustible. Such is the possibility that lies in front of you; for those who have got one glimpse of that, no earthly power has longer charm or desire. Before the radiance of that divine life, all glory of earth is poor and dim. This is not matter of faith, it is matter of knowledge; and every one whose vision is even partly opened will tell you that that only is the real life, and that the knowledge of the Divine is that which alone can satisfy the heart of Man.

Materialism Undermined by Science.

A Lecture delivered in 1895 in Calcutta.

IT is now fourteen months, my Brothers, since last I stood amongst you when I came to Calcutta last January twelve-month. I had only then made the acquaintance of, as I may say, the India of the South, with the various aspects that there may be found in her laws and in her religious thoughts. Leaving your capital city I travelled northwards and westwards and visited several parts of India, those of the North and North-west, and afterwards the Punjab. Thence I turned towards Bombay visiting several cities on the way, and then westwards back to Europe, there spending some months; and then southwards again to far Australia, where a new race is growing up, where a new nation, as it were, is being born: and from that far-off distant Isle, near to the South Pole, I come back once more to the Motherland amongst you again to bring you once more a message of the Eternal Verities of Spirituality, to speak amongst you once again the Eternal Truths which from ancient times have come down. For whether it be in India or Europe or Australia there is one mighty Spiritual Truth to be proclaimed, the one thing needed for the soul of man, and that is the knowledge of its wanderings after the Spirit, the knowledge of the Will of the Supreme. And whether in the lands of the West and South or whether under the fire of the tropical sun man is still demanding spiritual knowledge, is still struggling after spiritual life, still hoping for the same spiritual unity. To whatever land we may go, through whatever country we may pass, we have still Humanity as "the great orphan" crying for the Spirit, striving after Light, after spiritual unity, striving to find in the many exoteric religions the one Spiritual Truth which alone can satisfy the soul. And if I come back to you here and take up again the message which in this land has clothed itself in the ancient forms of Hindu religion from ancient times, it is not because India is the only land where human souls need it, it is not because India is the only country where the spirit of man is crying out for the Light, but it is because in this land there is more hope of a spiritual revival, and if a spiritual revival here there may be, then it will pour outwards to all the four corners of the world. For spirituality is more easily awakened in India than elsewhere. The spiritual heart here is only sleeping, whereas in some other lands it has scarcely yet come to the birth; for you must remember that in this land is the birth-place of every religion, and that from India, outwards, religions have made their way. Therefore it is that the soul of our mother India is so important for the future of the world, and therefore it is that the Materialism of India is so fatal. For it is here alone that lies the hope that man has of looking for spiritual life: for, in truth, unless the life of the Spirit come in this land, by reviving here, then the hope is baseless that spirituality is to spread over the world. And I may say to you, ere glancing for a moment over the subjects with which I am to deal, upon this visit, that in travelling through the length and breadth of India, from South to North, from West to East, I have found this of the people: that in the South of India you have more pronounced and outward orthodoxy, you have the more defined observances of ancient ceremonies and ancient rites, that on the surface of the people, as it were, you see more of the outer signs of Hinduism and more exactitude in the discharge of the various religious duties. That is a characteristic of the Southern people; that is a marked attribute amongst their various communities. Far away in the Punjab, there you may find certain traits of manhood, of strength, of courage, which if they shall rise to the Spirit surely would give us great help, would give us an enormous reinforcement; for that race would move with force and energy, only perhaps slow to take action. In Bengal there is, as I have noticed, much outward sign of western influence, much of the surface of the people taking up western thought and western customs; but in the heart of Bengal there still remain, more than elsewhere, gleams of the ancient spirituality, so that, just as in spiritual matters India is the heart of the world, so is Bengal the heart of India and may save India as a whole for all Humanity. And therefore in speaking to you in the ten days which lie before me, I have chosen subject after subject which should all point to the one object – and that is the revival of spirituality and the spread of the ancient Hindu religion in the hearts of its children, who are bound to it by ancestral ties. If you cannot revive spirituality in India through *Hinduism*, if you cannot thus reach India, then there is nothing else you can hope to do; and I say that here alone is the one hope of reviving this ancient potentiality. Here is the one certain hope which will bind all the hearts of the Indians into one and therefore we must look to the revival of the ancient faith – which however it has fallen, however much it has been corrupted in modern times, however much it may have lost spiritual life, is still the most ancient religion the world has ever known, sublime in its Philosophy and magnificent in its Literature. So that if this shall again become a living thing, India shall herself live; and with the revival all the sleeping truths of other religions shall look again towards their Indian mother, and make her once again the spiritual teacher of the world.

And now I am going to speak to you upon materialism; I am not going to deal now with a definite religious question, with definite religious teachings, with mighty doctrines in Philosophy, in Spiritual knowledge, which later on I shall hope to unfold before you. There is one thing that is eating the heart out of India, and that is modern materialism. There is one thing which is poisoning the mind of India, and that is the kind of science which is the teacher of materialism and works against Spirituality in the mind. How should I be able to tell you of the moral regeneration of India unless first I can strike at that which is piercing her heart and sucking out her very life-blood. So – as I have been trained in the science of the West, trained in the knowledge of the physical Universe which is so much used to make men believe that nothing but the physical remains – I take for my first subject his undermining of materialism by science, and I attack it with the weapons that were once used to build it up.

Now it is fair to ask in the beginning why it is that religion and science should appear to be in opposition. Why is it that science should seem to play into the hands of materialism? Why is it that as science has advanced, Religion has found itself pressed backward and backward so that men begin to make excuses for spiritual truths and talk apologetically of religion? Why is it that men advocating spiritual truth are afraid of being called superstitious? Let us see whether there is no explanation why science at the outset should help materialism and the reason also why, as science has advanced, it begins to undermine the same materialism and to destroy that which it has helped to establish? You may remember Bacon, a great philosopher of the 17th Century, speaking on this very point used the following phrase: – that a little learning inclineth men to atheism, but deeper knowledge brings them back to religion. It is a true statement. Look for a moment at religion and science, and you will see why that should be the fact, and why one should be against the other. A man who is a spiritual man – a religious teacher - regards the universe from the standpoint of the Spirit from which everything is seen as coming from the One. When he stands, as it were, in the centre, and he looks from the centre to the circumference, he stands at the point whence the force proceeds, and he judges of the force from that point of radiation and he sees it as one in its multitudinous workings, and knows the force is One; he sees it in its many divergencies, and he recognises it as one and the same thing throughout. Standing in the centre, in the Spirit, and looking outwards to the universe, he judges everything from the standpoint of the Divine Unity and sees every separate phenomenon, not as separate from the One but as the external expression of the one and the only Life. But science looks at the thing from the surface. It goes to the circumference of the universe and it sees a multiplicity of phenomena. It studies these separated things and studies them one by one. It takes up a manifestation and judges it;

it judges it apart; it looks at the many, not at the One; it looks at the diversity, not at the Unity, and sees everything from outside and not from within: it sees the external difference and the superficial portion while it sees not the One from which every thing proceeds. You may imagine, to take a figure, that you stand where there is a white light – say an electric light sending out rays from a single point; imagine three tubes going out from this centre and rays of light travelling down each and passing through a glass of a different colour set in each tube; if you look from the point where the electric light is you would see the white light striking outward as a light which was one; but if you went to the far end of the tubes you would there see that the light was of three different colours, as red and blue and yellow, appearing as if the light was of three kinds not one, because in their separation unity would be entirely lost. See how that works in the Universe. You have your three great *gunas* or attributes through which, as it were, the light comes as through three different glasses, and the one Divine Spirit comes down into manifestation; and it is not only the three *gunas* that you have but these intermingling one with another, and breaking in a thousand different channels. Then how great must be the differences at the circumference! But how it would lessen the difficulty if men could only see the processes, and know how those results were brought about; if they went further, and if travelling onward they found the divergences greatly diminish, see then how thus going forward, they may come, as it were, near to the one, and reconciliation between Religion and Science may arise. Religion shows everything from the point of the Spirit and proclaims the unity. Scientists show everything from the point of view of diversity and proclaim that, as if in opposition, to the world. But Plato says of the man who can discern the one in the many, that that man he regards as a God; the work of the true spiritual teacher is to show the one under the multiplicity, to make man see the fact of unity underneath diversity, and as science goes forward she also may be used once more to help us, because in passing out of the physical into the super-physical and mental, she is going nearer to Unity.

And now let me turn to my science and give you the proofs of this. First let me refer you, though I need not dwell upon the point, to the remarkable position taken by Huxley in his latest writings, which were new when I was with you last year, but which remain unchanged, uncontradicted, as the latest proclamation of the great teacher of Agnosticism as the latest proclamation of its exponent in European Science. Two great points he made, or rather three. First – and I only mention these briefly, because I dealt with them last year – first he pointed out that the evolution of virtue in man was directly in conflict with the evolution of the physical world: that when man evolved compassion, and tenderness and gentleness and self-sacrifice, when he learnt to use his strength for service instead of self-assertion – he was flying right in the face of the laws by which progress had been made in the physical Universe. He was following the law of self-sacrifice as against the law of self-assertion. Why is it that man can thus set himself against the cosmos? It is because he is approaching the spiritual region; it is because he has begun to develop the essential nature of the divinity itself: for the life of God is in giving and not in taking: the life of God is in pouring out and not in grasping; and as man feels the life of the Spirit in him against the life of the animal, he grows Divinely strong. And when you find men of science admitting that the evolution of virtue is by the law of self-sacrifice, you may perhaps begin to admit the possibilities of what is said in some of the sacred scriptures, that Creation always begins with Sacrifice. You may remember that - I am quoting to you, leaving out only the first great word – "the dawn is the head of the sacrificial horse, of the horse which arose out of the water, the water which the commentary says represents *Paramàtmà.*" All creation is Sacrifice. The source or dawn is the sacrifice, and everywhere the soul that would develop must live a life of sacrifice, because as the *Upanishad* says to you, a sacrifice of the Godhead was made in order that the world might exist. Sacrifice is the first condition in order that the Universe may be, and that man might be evolved to be one with Himself.

The second point made by Huxley seems taken from the sacred books of India; man can set himself against the cosmos because in man there is an intelligence which is the same as the Intelligence which pervades the Universe. That is the lesson of the Shastras. The intelligence of man is one with the Intelligence which pervades the whole. Man can set himself against the external world, for "Thou art Brahman," and when that is realised by man all else becomes subject to his will. And the third belief that Huxley has thought fit to declare is that the working of consciousness in the higher cannot be understood by the lower. There is nothing against the analogy of nature in supposing that there are grades of intelligence rising above men. There may be other intelligences higher and higher and higher, reaching further and further far above the noblest intelligence of man. And there is nothing, he says, to make it impossible that there should be in the universe, above these grades – a Single Intelligence. But what is that? Nothing but what has been proclaimed in the Scriptures, Isvar, the Lord, the Logos, the Word of which all things were made. So that you may see how, on these lines, science in the mouth of one of its greatest teachers is undermining materialism.

Now let me go a little further. Let us see, not from the mouth of the teacher, but from the facts themselves, how changes are going on. Physical facts are being discovered which show that underneath the material mind must be at work. Underlying the physical, intelligence must be active: underlying a particle of what was once called dead matter, a metal, a crystal or a stone, there is a moving life – there is a ruling intelligence. First let me say – and the force of the argument may excuse the repetition of it – that if you take a crystal, you find it grow along geometrical lines, with absolute definiteness of angles, as though a compass were used to trace it, and these lines make geometrical figures. So that Plato's phrase "God geometrises" is seen to be true even in the mineral kingdom. Then again when from the mineral you go to the vegetable where life is more active, where there seems to be less regularity, where there seems at first less of order, you will find in reality that even in its multiplicity there *is* order, that in the vegetable as well there is the same immutable law. If you take the branch of a tree, you may study the way the leaves are set, and you will find every leaf in a definite place, both as regards the leaves lower down and higher up. So that the leaves of the tree are developed on a geometrical plan. More than that. Since I last stood here to speak to you, a series of investigations has been made into the way that metals behave under exercise. Every engineer and other employer of machinery has noticed that when metal is used, where there are bars and wheels and other parts making up the machine, that with the use of the machine, what is called "fatigue" occurs. The metal gets tired. But what does this mean? It has been observed that, after a certain amount of exercise, the machine will not work well. It works like a tired horse or a tired man; it stumbles and cannot carry on the work. What shall be done? Let it rest. It does not want improvement, as every part is perfect; it does not want repair – there is nothing in it which is broken; it only needs to rest; and if it is allowed to rest it recovers from its fatigue, without a single thing being done to it, and it goes on to work as well as ever, showing that rest has given back its energies and that, just as a tired animal reposes, so also the "dead" metal may repose. This shows that even in a metal there is life – for a dead thing cannot get tired, a dead thing cannot lose its energies, a dead thing cannot be restored by rest. These are all signs of a living body; where there is fatigue and recovery of energies by rest, there is life existing, however hidden it may be under the form which conceals it from our eyes.

And now for a moment turn to Chemistry. I took first that point of the metals because it is a point which on thinking over you will find exceedingly plain and intelligible. But turn now to Chemistry. One great argument which materialists used to take from Chemistry was this: that as advances were made in what was called organic Chemistry, or the Chemistry of living things, it was shown that the separation made between organic and inorganic Chemistry was artificial. As a matter of fact, they said there was no fundamental difference and both organic and inorganic Chemistry were on the same lines; therefore they thought that the introduction of life as a thing separate and apart from chemical agencies must be given up. That argument was very much strengthened by chemists in the laboratory making certain things which before had been found only as products of vegetables and animals and which had been regarded therefore as the outcome of living energy. These things were said to be things which could only be produced by living organisations. During the present century, however, a large number of these bodies have been made by chemists, and they have succeeded here in breaking down the barriers between the organic and the inorganic; and the result was that at once it was said, "you see life is only, after all, the result of chemical energy, and not an outcome from the supreme source, but only something in connection with the chemical energy; you were under a mistake in supposing those things were always found as products of living things, and therefore there is not needed to explain them a source of life from which all living things proceeded. See how the chemist has proved you out of court; see how he has made that which you said could only come from life." Thus apparently was one of the arguments knocked down which seemed to prove the life of the world as coming from the life which was Eternal and Supreme. But Chemistry, in the course of these very investigations, going along the lines called organic, has given us an argument stronger than the one attacked. It places within our reach arguments far stronger, far more potent than the one which it destroyed; for it shows that in the organic the atom is not only, as I told you last year, formed by the action of electrical currents out of primary matter, but it shows further that the atom here progresses; that the atom in the mineral kingdom is not at all the same as the atom of the vegetable in its combining power. It shows that the change is not a change of material attributes, but a change of inner life, of internal differentiations - the atom changes within itself, as all living things do; for one of the great signs of life used to be said to be this power of adaptation from within. Take an atom in the mineral kingdom such as carbon. All its combinations are simple, all its combinations are one by one. This fourfold atom can join with others in definite and simple combinations, but when it passes forward, having gone through the mineral kingdom, then by an inner evolution, it changes its combining power and unites with itself to form a number of compounds, forming closed rings, so as to make complicated combinations never found in the mineral kingdom. Taking the old story of evolution as laid down thousands of years ago, not in the modern but in the ancient forms, we learn that this atom is part of the Universal life, that it is not dead matter but a living thing, that atoms are minute lives which go to build up external forms. We are able now to bring arguments from Chemistry to show that there is atomic evolution in the universe, that the progress of life which we see around us is no dream of the ancient Rishis but a reality. The scientists look only at the form and not at the inner life; but as you study the atom, you realise that this increased power of combination means evolving life within it. Not only is that seen, but it is also now admitted that life cannot be regarded as an outcome of chemical agency. It is admitted that life shows certain specific energies which differentiate it from electrical and chemical affinities, and you may get the phenomena of living things among the energies which science is unable to trace to their source. Once it was thought that life might be explained as the outcome of chemical and electrical agencies, but now it is admitted to be something more. Science now admits that although they are correlated with the life, they are not the life itself, and although they accompany the phenomena they cannot be regarded as their sources. So that from the chemistry which was the greatest hope of the materialist, we may now obtain arguments for its undermining.

Pass from that to electricity and see how here, in the latest discoveries, are arguments that may help our works. It is not only that science has proved that whenever thought is present, electricity is also present, interesting as that is, as showing the close relationship between them; but we are also told that there may be a development of an organ in the brain of man which will take cognizance of electric vibration directly and not indirectly. Let me show you what I mean. You see the light here because the light makes vibrations, and these vibrations strike on the organ we call the eye. The eye is so put together in its minute parts, that these vibrate in response to the vibrations of the ether; so that whenever these vibrations are present, certain particles in the eye vibrate in response, and give to us the sensation which we call light. Now these vibrations are within narrow limits;

there are vibrations in the ether both wider and narrower in wave-length than those which we call light, and to these our eyes do not answer. Therefore if they alone are present, we are in darkness; we cannot see. So again suppose we had developed the organ which is necessary to respond to the electric vibration, while we had not the organ of sight. Then this room would be dark to us, though filled with the vibrations we now call light. Then the consciousness could not perceive the light. But if we had developed instead of the eye another class of organs which answered to the electric vibrations, and suppose a large electric machine were fixed at one end of the hall, and a strong electric current sent through the hall, we should be able to perceive because the organ in us would vibrate in answer to the electric current, and the current would reach our consciousness through this organ. The consciousness is helpless without an organ that receives from without, and only the body can receive and transfer vibrations to the inner intelligence. That has been very clearly pointed out, and to take a striking illustration used by Professor Crookes: suppose we had no eyes to see the light, and suppose we had an inner organ which answered to electricity. This air would be opaque and we could not see through it, while a silver wire going through the air would be transparent, would be like a tube going through a solid mass. Though you would be able to perceive along the silver wire, because silver is a good conductor of electricity, you would perceive the air as a solid round the silver which would look like a hole. Do you see how rational the illusory theory can become when you learn a little more science? Do you see how matter is no longer the thing which it was, a solid material, but by a change in the organ of consciousness, what is solid to-day may be permeable to-morrow? And thus the idea is largely right that regards matter as an illusion; for what we call matter is only a generalisation of the impressions received by consciousness by way of the senses. It is the translation in consciousness of the unknown something which works upon us. In fact, what we call matter is but a reflection in the consciousness of an aspect of the Supreme Unknowable Unity, just as the Spirit is the reflection of the other aspect of the same Unknowable Unity. Thus science is bringing us back to this part of the ancient teachings, and if a materialist comes to you and says that matter cannot pass from matter, just throw into his mind for him to think over, some of these later facts.

Pass I from that to another closely allied point – that of thought-transference. Thought-transference is now being acknowledged, though for a long time science was very doubtful as to its acceptance, and if you spoke to a man about it he most likely regarded you as a crank, or even called you a fraud, for it was easier to call you a fraud than to admit that he was ignorant. There are men for whom it is impossible to say "I do not know," but anybody can say "you are a fraud." The ignorant who are not able to understand, people who are most self-opinionated nearly always call out "fraud," when confronted with the unintelligible. Look now at thought-transference. Thought is a form-producing force; when Brahma thought, worlds appeared. In the ancient books it was always taken to be granted that action is an effect of the mind. But it has been asked contemptuously of the writers of these books, what did they know about modern science? What did they know compared to our advancement? For we are supposed to know everything nearly in this 19th century! Yet, after all, the old writers have become justified by the facts. The old teachers have been justified by the later investigations. And some of the best of the younger scientists in England – the old ones are too prejudiced - are ready to take up facts, and they themselves have now performed the experiments that prove that thought-transference is possible. You have a man like Professor Lodge saying that his own experiments have convinced him and that he finds that thought can pass from mind to mind without what is called any material method. Not only he but the Psychical Research Society, which is an exceedingly "respectable" body from the public standpoint, have conducted a number of most careful investigations on thought-transference. The results of these were published in a book some three months ago by W. Podmore, a member of the Society. You will find in this book a record of most careful experiments on the transference of thought from one to another, and the evidence is now so strong that it is impossible to put it out of court.

Oliver Lodge speaking two years ago, said he was sure of thought-transference, but it was alleged that matter might be moved by the action of the will without material contact, and of that he was not yet convinced. But within the last few months Mr. Lodge has himself carried on a number of experiments which have convinced him, he says, beyond the possibility of doubt, that an article may be moved from one place to another without physical contact at all ; that bodies can be moved or suspended in the air without the means of physical support, and that he himself has taken part in experiments which have been carefully arranged by himself and other scientific men and they have proved that it is possible and it may be done over and over again. The experiments carried on included therein the taking of small articles and without physical contact passing them from one part of a room to another. The conditions under which these things were done were very rigid. They were carried on in a small island where there were no persons living except the lighthouse-keeper and his family. It was a very little island, a mere rock. Mr. Lodge and two or three others got the owner's consent to make their experiments there. They brought with them what is called a medium who belonged to the South of Europe, who could not talk the language of the inhabitants of the island, so that she could not communicate even with the family on the island, she being an absolute stranger never having been there before. They took her into a room with themselves only, with locked door, and there they performed the experiments in which these phenomena were produced. They kept the reporter outside in the balcony so that he could not be within sight of what was occurring. The reporter was put outside with a closed shutter between him and the people in the room. He was to write down what he heard, but he was not able to see what happened. Mr. Lodge said he was himself absolutely convinced ; he said he could not as yet explain it, but he thought it possible there might be a kind of expansion of vital energies by which a person, under certain conditions, could affect a body outside his physical reach. Just as one body can touch another by the exercise of physical energies, so can it draw others towards it. But he is not yet prepared to say how that energy is exercised. That this was, he knows ; how it was, he has not yet satisfied himself. But if he were to read some of the ancient books, he could easily find out. He might find that a man does not consist only of what is called the food-sheath or our physical body, but that men have other sheaths in which consciousness may work, without the limitations which are attached to the physical body. When it is working within there, it can also exercise its power, just as much as it can in the physical body, and may lift an object from one place to another by working with a law of nature in which other forces are concerned. The sheath used is what the Theosophists speak of as the "astral" body which can be utilised for the production of these phenomena, and though it was said to be a fraud when Madame Blavatsky brought an article from one side of a room to another, yet nearly four years after her death you have Mr. Lodge going into the subject, and asserting after a scientifically rigid repetition of the facts that the thing could be done, thus justifying a statement as possible which had been hastily dismissed as a fraud.

I might speak of many other cases of these latest investigations, and show you how they are undermining the materialistic idea. I may turn to Hypnotism, and remind you that last year I remarked that it was becoming a public danger – the power of influencing another, the power recognised by science, which one man had of imposing his thoughts on another. I saw that before long nations would be face to face with crimes which they would not know how to deal with. I said to you that unless the exercise of these powers were very carefully guarded, so that men who were unworthy should not be allowed to grasp these hidden powers of nature, there would be great danger to society in making safe particular classes of crime. Since last year that prophecy of mine has proved itself true, and in certain cases both in France and the United States of America crimes were found to have been worked by the hypnotiser, and the courts have not been able to deal with them, and verdicts of acquittal have been given on the ground that the criminals were not responsible for their actions, that being thrown into the hypnotised state, they could not justly be called to account by the law for the crime which they had committed. So that you have this result justifying the ancient practice of the East in withholding dangerous knowledge of occult forces, and showing that society in the West is face to face with the peril of men who commit crimes but who cannot be held responsible for them, because committing them under the influence of those who suggest them.

What is to be the outcome of these arguments? What is to be the outcome of these later investigations in Chemistry, electricity, thought-transference, Hypnotism, the moving of bodies and the like? To what are these new lines of investigation tending? They tend to show you that the old doctrine is true, that everything is the outcome of mind, that the Supreme Mind is, as it were, behind every phenomenon, that matter is regulated in conformity with the dictates of mind, that it is the truth that thought-forces take form in particular manifestations, and so the Universe is only an expression of the Divine Will. And inasmuch as the mind generates thoughts, and inasmuch as the Supreme and human minds are one in their essence, therefore the mind of man, in its higher manifestations shares in the powers of Supreme mind, and can control matter, can move matter, can model matter, shape matter, and make itself visible in the envelope of thought, and so communicate with other minds without any attempt to speak or hear at all. So that you begin to understand that the saying of the Puràna as to creation is not a dream, but that it is from the Supreme Will that forms emanate and build the Universe. And you may understand that this power of the Supreme is more manifest in the power of the mind than in the powers of the body, and that true activity is shown not in running about from place to place, held in the bonds of physical facts, but in quiet thinking, in the use of the imagination and the will. Therefore the Yogî sitting apart, with body absolutely still, with eyes closed, and mouth not communicating with other men, if he be a Yogî indeed, a Yogî in heart not only in dress, he has an inner life, a spiritual life, he may do more than the man of action by his thoughts, by his meditations, by the forces which are going out from him. On these more than on the work of politicians may turn the life of the nation.

Nor is this work only for the Yogi. Every one of you is sending out thoughts that, passing into the astral atmosphere, will take form, and thence affect the lives of men and in their totality the nation's future. If only every one of you would give one brief quarter of an hour's thought each morning to the future of India, and send out earnest wishes for her welfare, hopes for her revival, aspirations for her spiritual greatness, believe me you would make a force that would raise the nation and would mould her future. Your thoughts would gather together, modelling, as it were, an ideal India that should take shape in the external world; your prayers would gather together and ascend to the Feet of Mahádeva, whence would flow forth a regenerating energy that would manifest itself in teachers, in leaders, in guides of the people, who could move the hearts of men and unite them into one mighty Unity. Such is your power over the future, such the service you may render India: for in thought is the power of the Supreme, and it is man's because "Thou art Brahman."

Ancient and Modern Science.

From the "Theosophical Review," September, 1900.

IN these days of exultation over the advance of Modern Science, people are perhaps a little apt to forget that there still exists in the world a Science of vast antiquity, of hoary anciency, whose achievements dwarf those as yet obtained by its modern namesake, whatever of possibly overtopping greatness may await the latter in the future. It may not then be useless to study the two briefly, side by side, to see where they agree and where they differ, whether their methods coincide or whether they entirely diverge.

In speaking of Ancient Science, I am not thinking of the demi-antiquity of Greece or of Rome, the Science of which was merely an offshoot from that of Egypt, Chaldæa and India.

Ancient Science strikes its roots deeply into that vast continent over the greater part of which the waves of the Atlantic are now rolling, that continent of Atlantis which Modern Science is beginning to recognise, and the last fragment of which – the Island of Poseidonis, whereof Plato tells us – disappeared some eleven thousand years ago. Traces of that occidental Ancient Science are yet to be found in the records of Egypt and in the antiquities of China, and it is not without significance that the science of chemistry takes its name from Khem, the old name of Egypt. The Ancient Science that is more familiar to us is that which - brought eastwards by the flower of the Fourth Race that bore in its heart the seed of the Fifth – was planted in India and there grew into a mighty tree. While the Ancient Science of the West was whelmed under the floods, that of the East grew up in its stead, and in the first sub-race of the Aryan stock it was carried to a magnificent height and took on the sublimest developments. It is this which we will therefore take as the type of Ancient Science.

The point that at once strikes us when we first put Ancient and Modern Science side by side is the profound difference in their several attitudes towards Religion. In antiquity, Religion and Science were never divorced from each other, nor did it enter into the imagination of any to regard them as possible rivals. Every temple was a school; every priest was a teacher; and, for a reason that will presently be seen, a man needed to be a saint ere he could hope to be a sage. The Brâhmanas, the priestly caste, were also the teaching caste, and had it as their duty to train the young in all knowledge. And so highly was knowledge valued, that this teaching caste was the highest caste, and the ruler clad in cloth of gold would bow humbly at the feet of the half-naked but learned teacher, for it was thought a greater thing to add one small fragment to the area of knowledge than to bring another country within the confines of the empire. If Religion strove to reveal God to the heart, Science strove to reveal Him to the intelligence, and thus we find it written:

"Shaunaka, verily, the great householder, came near to Angiras full humbly and asked: 'What, O blessed one, is that which known makes known all else?'

"To him he spake: 'Two sciences should be known – thus the Brahma-knowers tell us – the higher and also the lower. Now the lower is the Rigveda, the Yajurveda, the Sámaveda. the Atharvaveda. prosody, rites, grammar, etymology, poetry, astronomy, and so on. But the higher is that by which the Eternal is understood.' "*

Within this Lower Science, the *Apara-vidyâ*, some four-and-sixty sciences were numbered, and for many patient years the student would strive for their mastery; but the Higher Science, the *Para-vidyâ*, that was but one, but a life-time could only learn its alphabet, for it was the crown of all sciences, the knowledge of the Heart of All, the Self. To know the Self, the Essence of nature, the Life universal, the supreme Being, the Eternal, that alone was knowledge – all else was ignorance. To know God was the last triumph of intelligence, the supreme achievement of Ancient Science, of the Science of the East.

Now the Science of the later West, Modern Science, strikes its roots in Southern Spain, in Andalusia, in the schools of the Moors and the Arabians. Fair fruit of the early days of Islâm, its very origin was an offence to the Christendom on which it was grafted. It came in the wake of invading conquering armies, and its presence was felt as a blasphemy against the Christ, as a triumph of His Mussulmân rival. The compasses were a weapon against the Faith like the scimitar,

*Mundakopanishad, i. 3–5.

and while the Muslim chivalry slew the body the Muslim university poisoned the soul. Religion seized, imprisoned, tortured, burned Science, and Science, forced to fight for its very life, for air to breathe, for ground to live on, struck with ever-growing force at the Religion that strove to slay it. Hence increasing antagonism, enlarging strife, the bitter "Conflict between Religion and Science," lasting down to our own days.

The difference between Ancient and Modern Science in their attitudes towards religion is thus due to the different environments in which they severally evolved.

The next point of difference that strikes us is that of the objects and line of study. Both work by observation, but the observation is directed along different lines.

Modern Science studies the forms that make up the kosmos; Ancient Science the life which holds it together and maintains each form. The first studies objects, and seeks by induction to discover the relations between them and the laws within which they act; the second studies the basic principles of the kosmos, and seeks by deduction to trace the path of evolution and to outline the necessary forms in which these principles will be expressed. It is as though in studying a tree one man began at the leaves, observed the shape, colour and characteristics of each, dissected them one by one, went from them to each twig, to each branch, to the trunk, to the root and the rootlets; the other took the seed, and, observing the life-principles at work, deduced their manifestations in root, trunk, branch, twig, leaf. The first studies the Many in its indefinite branches; the other the One in its indefinite expansion.

The order in which physiology and psychology are dealt with in relation to man will serve as a convenient illustration. Modern Science begins with physiology, studies the body, the nervous system, the brain, measures responses to stimuli, calculates the speed of nerve-waves, and so on, and on this basis proceeds to build up psychology. The individual consciousness is regarded as the outcome of all this nervous activity, and cannot be considered apart from it; to this conclusion this method of study inevitably tends.

Ancient Science begins with psychology, studies intelligence, analyses consciousness, investigates mental states, and regards the body as an instrument, an organ, shaped for the expression of these states. To it the body is *a result*, and consciousness can do without any particular body; let the one it is using be struck away, and it can readily fashion another.

The question at once arises in the mind: How can such a study be carried out? And the answer leads us to another profound difference between Ancient and Modern Science. When the modern scientist reaches the limits of his powers of observation, he proceeds to enlarge those limits by devising new instruments of increased delicacy; when the ancient scientist reached the limits of his powers of observation, he proceeded to enlarge them by evolving new capacities within himself. Where the one shapes matter into fresh forms, makes a more delicate balance, a finer lens, the other forced spirit to unfold new powers, and called on the Self to put forth increased energies. Why and how this was done shall be presently shown; that Self-evolution, or preferably that Self-manifestation, was the Secret of the East. Its first stages were in exoteric religion; its later stages in esoteric teachings. The end was the effectual shining forth of the Self omnipotent and omniscient, and when That was manifested all else became manifest afterwards.

Before dealing further with this let us glance at some results of modern study, which have carried Modern Science into a field whereon it meets its ancient predecessor. The common ground on which this meeting takes place is the ether. The two start from opposite poles, and meet at last here. Modern Science has climbed slowly upwards, making sure each step of the ascent; solids, liquids, gases, have been observed, weighed, tested, analysed, and at last Science finds itself in a region where matter fails to respond, becomes intangible, imponderable, and yet it must be present to render intelligible the working of mighty energies. So Science formulates the existence of intangible, imponderable matter – intangible, imponderable for its present resources – and proceeds to study it as best it may. Ancient Science has descended step by step from life and intelligence to the kinds of matter in which they clothe themselves, becoming ever denser and denser, till it also reaches the ether and carries on therein its later observations. Here, then, we can compare their results, and see how far they agree.

Among the more significant of late discoveries has been that of the Röntgen or X-rays, vibrations in the ether which pass through matter hitherto regarded as opaque, and, for instance, enable a photograph to be taken of the skeleton within a living body, or of a bullet imbedded in an internal organ. These vibrations are alleged to be seventy-five times smaller than the smallest light vibrations, and thus can pass through matter impermeable to light and heat. Now eight years before the X-rays were discovered *The Secret Doctrine* was published, and in that Mme. Blavatsky remarked : "Matter has extension, colour, motion (molecular motion), taste and smell, corresponding to the existing senses of man, and the next characteristic it develops – let us call it for the moment 'Permeability' – will correspond to the next sense of man, which we may call 'normal clairvoyance.' . . . A partial familiarity with the characteristic of matter – permeability – which should be developed concurrently with the sixth sense, may be expected to develop at the proper period in this Round. But with the next Element added to our resources in the next Round, Permeability will become so manifest a characteristic of matter that the densest forms of this Round will seem to man's perceptions as obstructive to him as a thick fog, and no more." * The fulfilment of the latter part of this quotation lies in the future, but the earlier part is now verified, for the discovery of the X-rays has completed a singular chain. Not long ago, a little boy in America saw the bones

* *Op. cit.*, i. 272, 278, last edition.

of his father's hand through the covering flesh, and medical observations established the fact that he "saw by the X-rays," or, to use our own phrase, was "physically clairvoyant." Other people here and there show this faculty, born with them, "variations" pointing to a line of evolution. Under hypnotic conditions many persons show this same power, and "hypnotic lucidity" is a well-established fact. Others become clairvoyant by practice. Surely when these facts are set side by side: etheric vibrations by which certain objects may be seen through opaque matter; occasional instances of people born with a power to receive and respond to those vibrations; many people able to receive and respond to them when shut off from the vibrations they normally respond to; artificial development of the power to receive and respond to them; we have definite signs of the evolution of a new sense and sense-organ. The sense-organ is rudimentary in the normal person, is at least partially developed in the born clairvoyant, is susceptible of stimulation in most people when the developed senses are temporarily silenced, and may have its development forced by special means. Here the positive declaration of Ancient Science, based on innumerable experiences, is in way of verification by the discoveries of Modern Science.

The existence of what occultists from immemorial antiquity have called "dark light," or "invisible light," is being proved by the experiments of Dr. Le Bon, related by himself in a monograph, of which the salient points are quoted in the Parisian *La Nature* for June, 1900. Conscious or unconscious of its significance, he has named his discovery *la lumière invisible.*

Ancient Science asserts that etheric vibrations can be utilised for purposes of communication without the employment of apparatus connecting the points of generation and reception. Jagadish Chandra Bose and Marconi have severally proved this to be true as regards some such vibrations marked off as electrical. "Wireless telegraphy" is now an established fact, and shows that the ether itself suffices as a medium of communication between widely separated points. The transmission of thought-waves through the ether is thus proved to be theoretically possible, and its actuality is asserted by such eminent scientists as Sir William Crookes and Professor Oliver Lodge, to say nothing of less important investigators.

Another interesting statement, made by

Marconi, may be mentioned : that he believed that the electrical vibrations were of different *forms*. Herein he is quite at one with Ancient Science. Some observers, who study according to the old rules, have stated that the form of the X-ray vibrations is a double spiral or helix. It will be interesting to see if any later scientific discovery verifies this observation.

As Modern Science continues its discoveries in the etheric region, it will more and more substantiate the assertions of Ancient Science, reached by methods so different from its own.

As the limits of our space forbid us to further multiply instances of concord between Ancient and Modern Science in the etheric region, we must turn to the question already formulated : Why did Ancient Science begin with consciousness, and how can study be carried on along its lines ? Thus is it answered :

The universe consists of the vibrations of a universal life, and of the forms into which they throw the matter in which they play. Life is motion. Consciousness is motion. Forms vibrate under its impulse according to the rarity or density of the matter of which they are composed. The life vibrating within a form enters into relations with, affects, any other portions of life within forms which are capable of responding to it, *i.e.*, of reproducing its vibrations in whole or in part. At a certain stage of this exchange the separated lives become conscious of each other.

The Self in man is part of the kosmic Self, and is capable of vibrating in every way in which the kosmic Self vibrates. This Self in man is the "I" which is conscious of its own existence, which feels and thinks. As it exchanges vibrations with other Selves around it, it distinguishes all in which it is not conscious of its own existence, in which it does not feel and think, as the Not-Self. (The separation of forms leads it to the false conclusion that the Selves are also separate.) This Self can only know the other Selves as it is able to respond to them, and its "evolution" is merely the bringing out of the capabilities it contains. Hence it can know everything by turning outwards the powers within it, and all true knowledge is attainable by Self-unfoldment only. We know a thing when we become it, *i.e.*, when we vibrate as it vibrates. The bodies with which the Self is clothed enable it to come into touch with all bodies composed of similar materials, which vibrate at the same rates.

In the present solar system there are seven fundamental types of matter, elements or atoms, primitive bases of all combinations. Each of these types gives rise to innumerable combinations, which in their totality form a "world," or "plane," or "region of existence." The Self clothes itself in a body or sheath of each kind of matter, and thus comes into touch with all these worlds, each body receiving and responding to the vibrations of its world. Consciousness is the relation between the Self and the Not-Self, and the expansion of this relation is evolution. As the physical world is known by means of the physical body, through which the Self receives it, so each world of lessening density is perceived by the Self through a body of similar matter. Further, these bodies are separable from each other, and the Self can temporarily discard the grosser to facilitate its observations of the subtler.

The fundamental principles of Ancient Science were established by the experiences of highly-developed men, and are always verifiable anew by those who develop the capacities inherent alike in all. But this development is, it is fair to say, not practicable for everyone within the limits of the present life, any more than great scientific attainments can be said to be within reach of the majority. If a man is to become a great mathematician, a great astronomer, a great physicist, he must begin life with a marked aptitude for the branch of science in which he is to excel. To this marked aptitude he must add careful and prolonged study, aided in the earlier stages by competent instructors; he must give his life to his work, if he is to achieve eminence, and must make other pursuits subordinate to the one aim of his life. All this is equally necessary for the man who would win success in the pursuit of the highest Science, and first-hand knowledge cannot be enjoyed by any who do not fulfil these conditions of all successful pursuit of truth in any kingdom of nature.

It will now be clear why religion could not be divorced from Science among those who thus regarded life. The first thing necessarily demanded from the student was that he should cease from evil ways and dominate his passions, so that the Self might utilise its lower vehicles for the gaining of knowledge, undisturbed by riotous vibrations which blurred all vision. Then the student was taught to refine the physical body and render it ever increasingly sensitive to vibrations, while preserving it in health. He was trained to control his senses and to concentrate his mind, until, having purified and thoroughly mastered his vehicles, he could use them only for the purposes of the Self. Then he could come into touch with every part of nature, and for such a one "Nature has no secrets in all her kingdoms."

Along such road travelled Ancient Science, and for those who would still follow that Science there is no other road.

Those who are not yet prepared to tread this ancient path, may yet do much to profit by the suggestions and hints gathered from Ancient Science, if they will avoid falling into extremes in Religion on the one side and in Modern Science on the other.

There are two great enemies that ever stand opposed to human progress, one wearing the mask of Science, the other the mask of Religion. One is the Incredulity which denies facts because they are new; the other is the Credulity which accepts superstitions because they are old. Each grasps a poniard with which it strikes at Truth, the Angel which guides Humanity along the upward path. Which is the more dangerous foe it is hard to decide, for the rigid refusal to even consider the evidences on which a new and startling truth reposes, shuts a man out from progress as much as does the folly which swallows open-mouthed the emptiest tale. Superstition often renders a man more ridiculous than does scepticism, but their effects on progress are much the same. Hard iron cannot be shaped any more than fluid mud. We need willingness to study, impartiality, clear vision and right judgment, and then we shall find that now, as of old, right knowledge is attainable, for we have within us That whose nature is knowledge, and who can never rest until He can say "I know."

Modern Science and the Higher Self.

A Lecture delivered in 1904 in India.

THE putting together of the two phrases — "Modern Science" on the one side, and the "Higher Self" on the other, may strike some of you as strange and even as incongruous; for the ideas of Modern Science and of the Higher Self are so far removed from each other in the general mind that to bring them together as though they were closely related must seem to be unusual and grotesque. And yet I think I shall be able to show you as we go on, that these two things, the most modern and the most ancient, the thought of the West working by way of continuous experiment, and the thought of the East working by seeking the

Higher Consciousness and recording its testimony, that these two are in our own days brought closely into touch with each other, so that they may aid and strengthen each other, may be found as servants in a common cause, and not as opposing and incongruous ideas. I want to show you, in the course of this evening's lecture, that there is in Modern Science a distinct recognition of the Higher Self, that there is an agreement between eastern and western science, conflicting with each other in their methods, that there is a mass of evidence compiled by western scientific men, which can be cited as showing the recognition by Science of the Higher Self, of the existence of a Jīvātmā, a living Spirit, a living intelligence in man, and that the Spirit finds an ever imperfect instrument for expressing itself in the body of man. I want to show you how the face of Modern Science to-day is turned in a different direction from that in which it was turned some 20 or 30 years ago, I want to show you that there is a growing idea in the West, that man in the waking consciousness is but a small fragment of the real man, that man transcends his body, that man is greater than his waking mind and consciousness, that there is evidence in plenty, daily forthcoming from most unexpected quarters, to show that human consciousness is far larger and fuller than the consciousness expressed through the physical brain. This idea of a larger consciousness, larger than the normal waking consciousness in man, the consciousness hitherto recognised in modern psychology, is one that has not only been suggested but is now beginning to be recognised by Modern Science in the West. Such is then the reason for putting these two phrases together "Modern Science" and the "Higher Self."

Now, I ought to define what I mean by the "Higher Self." I am not using the phrase in the strictly technical sense which you find in the Theosophical literature, that is to say, the Jîvâtmâ in man. I am using it for the whole expression of that Jîvâtmâ above the physical. I am using it for everything which transcends the brain consciousness, which finds the brain too coarse and dense an instrument for its expression. I am using it, in short, to imply what generally goes under the term "larger consciousness." If we can show definitely that experimental science has recognised human consciousness to be stronger, more energetic, more lively than the consciousness working in the physical brain, if we can prove the existence of yet higher realms, we shall enter on a path which leads to the highest invisible worlds. We climb step by step and see the larger consciousness unfolding itself more and more, stretching over an immense expanse, till at last we reach that, to which men in every clime have always aspired, till the spiritual aspirations of man are vindicated. Such is the promise of infinite expansion which lies in the domain of an enquiry into the consciousness of man. The particular branch of Modern Science which thus comes into touch with the Ancient Science is that of psychology. Psychology in its modern form, climbing from below by way of experiment comes into touch with the ancient psychology of the East ; and this is a science of immemorial antiquity, whereas modern psychology is an infant science in the West. Not that the West had no psychology ; in the Middle Ages and in the centuries that went before them there was a psychology, but that psychology was repudiated in modern days. So that if you go back some thirty or some five and thirty years, you will find it distinctly stated by the representative European thinkers that no psychology could be regarded as sane which was not based on the science of physiology.

The method of introspection, of the observation of one's own mental processes, was entirely discarded in modern thought. The method of studying the mental processes of others by inference was equally challenged and doubted. It was said, and there was some truth in the saying, that the moment you began to study your own mental process, that moment it changed by the very fact of your observation; and it was argued also that if you tried to study the mental process of others, you could only do it by inference and not by direct observations. It is necessary, it was said by modern thinkers, to first study the brain, the nervous system and mechanism in man, whereby thought is expressed. Thus arose the science which is called psycho-physiology, a science in which the nervous system and the brain, regarded physiologically, were studied, were analysed, were measured, and it was considered impossible to understand thought without the knowledge of thought's mechanism, and without a knowledge of the process of the changes in that mechanism. It was considered that along this road of experiment better results would be obtained than would be obtained by other methods, and as science became more and more materialistic, as it reached the point at which Professor Tyndall made his famous statement that we were to look in matter for the promise and potency of every form of life, it was natural, if not inevitable, that men should begin to study thought by the study of its mechanism, of its apparatus, rather than by the way of the direct observation of its processes. As the method of experiment justified itself more and more by most interesting results, it became regarded as the only method which was worth the consideration of the thoughtful, of those untainted by superstition, hence the birth of what may be called a new science, the science of psychology based on physiological experiments, a science which it was thought would confirm the statement that thought was the product of the brain, was really the outcome of the nervous mechanism in man.

So far were men inclined to go in making rash statements, that it was deliberately laid down that thought was produced by the brain. You had such a well-known, such a famous physiologist as the German Carl Vogt, declaring that the brain produced thought as the liver produced bile. That, perhaps, was the most extreme statement of the school of thought represented by many of the German thinkers. The very fact that such a statement could be made showed the line of thinking which Modern Science was following.

Now directly in opposition to that stood

the immemorial psychology of the East. That was founded on the idea that man in his essence was not a body but a living Spirit, was not a mere form but an eternal Intelligence. Eastern psychology was founded on the notion that this living Intelligence, this entity, Jiva or Jivàtmâ, was the primary thing to be understood, that instead of considering thought to be the product of a certain arrangement of matter, the certain arrangement of matter was to be regarded as the result of thought. Instead of considering that life, consciousness, intelligence were the results of a mechanism, of an apparatus gradually built up by nature under the working of blind and unconscious laws, eastern psychology declared that the primary fact was the fact of consciousness and that matter was but its garment, its instrument, the means of its expression, arranged and guided by intelligence, and only useful and interesting as expressing that intelligence in the various worlds of the universe. That is put strongly and clearly in the *Chhândogyopanishat,* and I quote this because we shall find in the latest science a strange and startling endorsement of the ancient thought. It is declared in that Upanishat that Atmâ exists, and that the bodies and the senses are all the results of the will of Atmâ. You may remember how the passage runs: "The eyes are for the perceiving of that Being who dwelleth within the eyes." It was Atmâ who desired to hear and to smell and to think; hence arose the organs of the senses and the mind. Everywhere Atmâ is the primary fact; the organs, the bodies, come into form in order that the will of the consciousness may be expressed. Thus great, then, is the opposition between this western thought of some thirty years ago and the ancient thought of the East, the one beginning with the body out of which the consciousness grows, the other beginning with the consciousness by the activity of which gradually the bodies were formed. Keep this antithesis in mind as we follow out the lines of our study.

Come then to Modern Science, starting with the idea that thought must be understood by the clear understanding of its mechanism which many considered its producer. Great scientists began to study carefully the nervous system, and they studied it with a marvellous patience, they studied it with marvellous success; they measured the rate at which the impressions made on the surface of the body travelled to the nerve centres and there appeared as mental perceptions. They measured the rate at which the thought could travel along the nervous fibres. They measured the delicacy of perception related to the various parts of the mechanism. They reduced more and more all thought-processes to processes of chemistry, of electricity, to be measured by the balances, to be weighed, to be analysed. They met with great success; they threw wonderful light on the mechanism of nervous apparatus. They went, in their researches, in their efforts to map out the nervous system, even into crime, the crime of vivisection. Thousands of miserable animals had their skulls taken off, their brains laid bare, and electrical shocks applied to the various parts of the brain, in order that by these diabolical methods some of the secrets of consciousness might be wrenched from Nature.

But as they carried on their experiments, certain results appeared which seemed to challenge the starting point from which they had come. They were dealing with thought as the product of the nervous system, and necessarily, therefore, they thought as the nervous system increased in perfection, the thought increased in complexity, in accuracy, in dependability. The constitution of the brain, the relation of the parts of the brain to one another, the functions that belonged to different portions of the brain, all these were mapped out, analysed, explained.

But as they began to study, or rather as they carried on the study, they found that there were certain results of consciousness that did not fit into the theory with which they had started. They found that there were certain results of consciousness which took place when the brain was not in its normal state, in its full activity, and that these could not be ignored, that no theory of consciousness could be true that did not explain these as well.

First the attention was turned to what were called the results of dream consciousness. The waking consciousness had been carefully examined. But what of the consciousness that went on when the man was asleep? The phenomena of sleep must also be explained. Interesting experiments began on the dreaming consciousness, on the functioning of consciousness when the body was asleep. Experiments were made with the usual care, with the usual ingenuity, with the usual patience. We have no time to deal in detail with them. Many of you can find for yourself a large amount of details in the famous book of Du Prel, *The Philosophy of Mysticism.* It seems sufficient for the moment to give you one example which covers a large class of experiments. They began by taking a sleeping man, touching his body at some point, and then waking him at the moment, and asking him : " What have you dreamed ? " Very often there was no dream, no result, but in a large number of cases the man reported a dream. I will take one illustration. The back of the neck was touched. The man was asked : " What have you dreamed ? " He said : " I dreamed that I committed a murder ; I was brought to trial for the murder ; I dreamed the whole of the trial. The speeches of the barristers, the summing-up of the judge ; I waited for the verdict of the jury ; I was pronounced guilty ; I was doomed to death ; I was taken away to the condemned cell ; I remained there for so many days ; I was led to the place of execution and, as the knife of the guillotine descended on me, I felt it touch my neck, and I awoke." Many such experiments were tried and put on record. What was the result that came out of them ? The stimulus to the dream came from the touch, as the touch on the back of the neck had suggested the idea of death by guillotine. How can we explain all that went on in the dream consciousness after the touch and before the awakening ? How did the dream consciousness pass through a long series of events in order to explain the touch, and how was it that the events which followed the touch seemed to precede and to explain it ? That was the problem.

After long discussion and cogitation the suggestion was made that consciousness in the dream state was working in some medium other than the dense matter of the cells of the brain. The speed by which the nervous impulse travelled from any part of the body to the brain had been measured and was known. But here was a long series of phenomena in consciousness which came between the touch on the body and the knowledge of that touch in the brain. There must therefore be some finer medium in which consciousness was working, through which the knowledge had gone more rapidly than through the nervous matter, so that there was time to build up the story to account for the touch before the consciousness of the brain knew it had been made on the body. The mind then, in sleep, was not thinking by the dense brain, but by some subtler medium which answers more rapidly to the vibrations of consciousness, just as two men might start from the same point and one running quickly might run to the goal fast, turn back, and might meet the other long before he had covered a quarter of the distance, so consciousness in the subtler medium could travel faster, make up the story to explain, return and meet the consciousness in the physical brain, and give the story as the explanation of the touch that had been felt outside. Such was a suggestion made to explain the rapidity of action in the dream consciousness. But much more than that was wanted to make a satisfactory theory. And science said: it is not enough to have dreams examined in this way. Let us try to throw a man into the dream-state and examine him while he is in it instead of waking him out of it. Let us try to come into touch with him while the dream is going on. Then began the long series of experiments spoken of as hypnotic, where a man was thrown into a trance and thus a prolonged dream-state was attained, a state produced by and under the control of the operator. In the hypnotic trance, as you must know, the body is reduced to the lowest point of vitality. The eye cannot see, the ear cannot hear; lift up the eye-lids and flash an electric light into the eye, there is no contraction of the pupil; the heart well-nigh ceases to beat, the lungs have no perceptible action; only by the most delicate apparatus can it be shown that the heart is not entirely still, that the lungs have not entirely ceased to contract and to expand. Now what would be the state of the brain under these conditions according to the theory that thought is produced by the brain? The brain is reduced to the condition of coma, lethargy. It cannot work; it is badly supplied with blood, the blood it obtains is surcharged with carbonic acid and waste products, for it has not been supplied with sufficient oxygen. From such a brain, according to the modern theory, no thought ought to be able to proceed. But what were the startling facts that answered the enquirers, when they questioned the consciousness under these abnormal conditions? Where there should have been lethargy there was increased rapidity, where there should have been stupor there was very much increased intelligence, where there should have been apparent death, there was life in overflowing measure, and the whole of the mental faculties were stronger and more vivid.

Take the memory of man in the waking state. Question him about his childhood, he will have forgotten many events, they have vanished into the past. Throw the man into a trance, question him then about his childhood, and memory gives up the stores that apparently had been lost, and the most trivial incidents are recalled. Take a man, read to him in his waking moments a page of a book that he has not before heard, and ask him to repeat it; he will stumble over a sentence or two, he is unable to recite it. Throw the same man into the trance, read the page to him then, and he will repeat it word for word even when the language is to him unknown. Memory then in the dream or trance state is immensely increased in its range and power. Take the perceptive faculties. As I said, the eye would not answer to the flash of the electric light, but the faculty for perceiving the material world of which the eye is the organ finds expression in the trance state such as in its waking state it cannot exert. The man in the trance will be able to see through a barrier that blocks the waking vision, and tell you what is happening on the other side of a closed door, or what is within the body, tell you not only what is happening on the other side of an obstacle but what is happening hundreds of miles away. These are not the dreams of the orientals, of the theosophists, I am confining myself to cases where experiments have been put on record, where men who do not believe in the superphysical, were confronted by facts they found it impossible to explain. I have myself seen experiments of this kind in the days before I was a theosophist. These proved to demonstration that opaque bodies were not obstacles to the vision of the man plunged into the hypnotic trance, and this is now admitted by all students of hypnotism. Memory and perception then are increased in power when the brain is stupefied. So I might take you through one faculty of the mind after another and show you that in every case consciousness is stronger, more vivid, more active, when its physical mechanism is paralysed.

Out of all these experiments there arose again the question: What then is the relation between consciousness and the brain? It was established that with paralysis of the brain consciousness becomes more active than it was before.

The result of these experiments on the condition of mental faculties, was a proof that whatever the dream consciousness of man might be, it was far wider in extent, far more powerful, than the same consciousness working through the physical brain. Thus gradually way was made for the recognition of the fact, well-known to the eastern psychologists, that the waking consciousness is only a part, an imperfect and fragmentary expression, of the total consciousness of man.

The modern psychologists meanwhile were proceeding on a new line of investigation, and they began to study what are called the abnormal phenomena of consciousness, not only the normal and the common-place but the abnormal and the exceptional, and at first the study along these lines seemed to carry most of the thinkers directly against the psychology of the East. Study in one school of psychology came to what seemed a terrible conclusion. It was the school of Lombroso in Italy. He declared, and many others followed him, that the visions of the prophets, of the saints, of the seers, all their testimony to the existence of superphysical worlds, were the products of disordered brains, of diseased or over-strained nervous apparatus. He went further, and he declared that the manifestation known as genius was closely allied to insanity, that the brain of the genius and the brain of the madman were akin, until the phrase "genius is allied to madness," became the stock axiom of that school. This appeared to be the final death-blow dealt by materialism to the hope of humanity nourished by the grandest inspirations that had come to men through the geniuses, the saints, the prophets, the seers, the religious teachers of the world. Was all this truly but the result of disordered intelligence? Was all religion but the grand dream of diseased brains and nerves? Was religion really a nervous disease, are all people who see and hear, where other people are blind and deaf, neuropaths? That became the terrible question set rolling by this psychological school. At first there was silence, caused by the very shock of the question. Men were so taken aback that they knew not how to answer them, knew not how to argue. Gradually, however, there came from the ranks of thoughtful men a challenge. Granted that this is true that you have discovered, granted that these brains through which the visions of religion, the revelations of religion, have come, are abnormal, is it after all so important, so vital a matter? May it not be that as the higher worlds come into touch with man, they may well be able to affect only the most delicate brains, and in that very touch they may throw the delicate mechanism slightly out of gear? Is it not possible that the subtle vibrations of the higher worlds to which the human brain is unaccustomed yet to respond, may in some individual differing from the standard of ordinary evolution, find an answer, and the higher world may speak through these abnormal brains to men? The question of importance to humanity is not whether the physical brain of the genius is allied in its mechanism to the physical brain of the madman, but whether what comes through the brain of a madman and the brain of a genius are equally important to humanity. If we receive through the disordered brain of a mad man, a jangle of useless disconnected ideas and dreams, that result is worthless and we set it aside. But if through the brain of the genius, of the religious teacher, through the brain of a prophet, through the brain of the saint, come forth the highest inspirations, the loftiest ideas that have raised mankind above the brute and the savage, shall we cast them aside as well? We judge the results not by the mechanism through which we have received them, but by their value to humanity, and no matter what the mechanism of the brain may have been there remains the thought that has been given to the world.

Everything of which humanity is most proud, all its sublimest hopes and aspirations, the most beautiful imaginings of poetry, the transcendental flights of metaphysics, and the sublimest conceptions of art, are all product of neuropaths, of abnormal brains. When men tell us that the great religious teachers are neuropaths, that Buddha, Christ, S. Francis, are neuropaths, then we are inclined to cast our lot with the abnormal few, rather than with the normal many. We know what they were. They were men who saw far more and knew far more than we;

what matters it whether we call their brains normal or abnormal? In these men's consciousness is a ray of the Divine splendour; as Browning says:

> Through such souls alone
> God, stooping, shows sufficient of His Light
> For us in the dark to rise by.

And if in those cases the brain change from a normal to an abnormal state, then humanity must ever remain thankful to abnormality.

That was the first answer which may be made to this statement of Lombroso, and you find a man like Dr. Maudsley, the famous doctor, asking whether there is any law that nature shall use only for her purposes what we call the perfect brains? May it not be that for her higher performances she needs brains which are different from the ordinary, the normal brains of man? For take the normal brain that is the product of evolution, the result of the past; that brain is fitted for the ordinary affairs of life, it is fitted for the calculations of the market-place, for the observation of material things, for the work of the world, for the carrying on the ordinary affairs of life; brought to its present state by the practice of such thinking it is the best machinery for such work. But when you come to deal with higher thought, with abstract speculation, when you come to deal with religious ideas and with the possibilities of higher worlds, that brain is the most unfit of instruments ; it is not delicately organised enough for the subtler vibrations of the higher worlds. For just as you may take a watch delicately wrought and by that watch you may measure small intervals of time which you could not measure by a clumsier mechanism, so it is with the different brains of men. The normal brain is the common-place brain ; the normal brain is the average brain. It has no promise for the future ; it is but the product of the past. But the abnormal brain, that which can answer the higher vibrations, the brain which, if you will, you may call by the insulting name of "morbid," that is the brain which stands in the front of evolution, which is the promise of the future, and shows us what man shall be in the generations yet to come.

As the struggle went on another answer came. When in times of unusual strain and unusual excitement, the brain answers to the higher vibrations, then it is very likely that nervous disease will accompany the answer. It is not always the brain of the genius to which strange experiences occur. They occur to people of all types ; the average man and woman have their experiences. When a person has been rapt in ecstasy of prayer, or is fasting, and the body is weakened or is under great stress of excitement, the brain will certainly be affected far more easily than under normal conditions, and it will be able to register finer vibrations more easily than the so-called healthy brain. Take a very common illustration. You have a violin or a vînâ. You find that you can get from the string of your instrument a certain note, but you want a higher note ; what do you do ? You tighten the string. Just so with the human brain. It does not answer to the higher notes of life in its ordinary state ; you must tighten the string by intense concentration or devotion, and then the brain will answer. But in the tightening there is danger ; in the tightening there is possibility of breakage ; and so in the normal brain tightened to respond to the stress of the subtler vibrations, there is the danger of nervous disturbance, there is the danger of unbalancing the mechanism.

How can that be met ? We look to the science of the East, to its old psychology, and it gives us the answer - it is the only science that knows the answer, and this answer is strengthened by a modern discovery touching the mechanism of the brain. What is the process of Yoga? It is a process by which gradually, by physical, by mental training, the man develops a higher consciousness, and enables that higher consciousness to express itself in the physical body. Now every one of you who knows anything of Yoga, knows perfectly that in Yoga there is a physical training, purification of the body, purification of the brain, which precedes the practice of any of the higher forms. You know that it has always been told that if a man would practise Yoga he must become an ascetic in his life. That he must give up liquor and the grosser articles of diet, that he must purify the body, and then purify the mind. You know that only as that is being done, can the mental Yoga be effectually performed, and then as the body is becoming purer day by day, consciousness develops, its higher powers show themselves through the purified brain without disease, without over-strain, without any injurious nervous or morbid results. Eastern psychology recognises the danger of nervous disturbance, and enables the necessary sensitiveness to be obtained without the overstraining of the physical instrument.

But I said that a late discovery in Modern Science with regard to the brain had justified the process. What is the discovery? That the brain cells in which thought is carried on develop, and increase in size and in complexity by the process of thought; that as a man thinks, his brain cells grow; they send out processes which anastomose, join one to another, and thus make a very complicated mechanism by which higher thought can be expressed, that the whole process of the expression of thought depends on this growth in the cells of the brain, and that as you think you are really making your brain, you are creating the mechanism by which hereafter a higher thought may be expressed. The latest anatomy of the West has laid down this, that these cells grow under the direct impulses of thought, and that as you think you prepare the brain for better and higher thought; as the thought acts on the nervous cells, the nervous cells become more complicated, intercommunicate more fully, are more apt for the processes of thought.

The Yoga practice of concentration, of steadying the mind by fixing the thought, makes the brain cells grow, and thus creates an instrument adaptable for higher thinking in the future. As you carry on your meditation you are building fresh mechanism in the brain; as you carry on concentration, you are creating the apparatus for higher performances. Thus as purification of the body, of the brain, of the mind, goes on in Yoga, you are building up the brain, making it able to come into touch with a higher world, without losing its balance, without losing its sanity and its strength. There lies the scientific justification of Yoga from the latest investigations of the West. What then does the East tell us as the result of Yoga? It tells us that man is a consciousness expressing his powers through the body which he moulds to his own purpose, that man's consciousness in the brain is far less than his consciousness out of the brain. That man uses the brain as an instrument on the physical plane, but is not limited by it, is not confined by it. That old theory of the ancient Sages is now being promulgated in the West by such men as Sir Oliver Lodge, who declares that the investigation of hypnotism, the study of consciousness, the study of abnormal states of consciousness, prove that human consciousness is larger than the consciousness in the brain, and that there is much more of us outside the body than is shown by the working of the brain. That is the last word on consciousness from the West, and it is identical with the testimony of the East.

Do you see now why I put together Modern Science and the Higher Self? The Higher

Self is the consciousness beyond the physical, the larger, wider, greater consciousness which is our real Self, the Self of which the consciousness in the brain is only the faintest of reflections. This body of ours is only a house in which we dwell for our physical work; we hold the key of the body; we should put it in the lock by Yoga, and try and release the imprisoned consciousness. We are greater than we appear to be; we are formed in the divine image: we live not in this world only but also in other worlds; our consciousness outstretches the physical. In this planet of mud our foot is planted, but our heads touch the heavens; they are bathed in the light of the spiritual world far above, in the world unseen, bathed in the light of our God. We may trust the consciousness and the testimony of the Saints, the Prophets. the Seers, and the Teachers of humanity. They told us what they knew, that which we may also know for ourselves. They were divine, showing Their divinity to the worlds. We are none the less divine, although our divinity is veiled. Let us claim our birthright, to know as They knew. These great ones of the past, these Saints and Teachers of humanity, They are the promise of what we shall be in the future, and the heights They have touched in ages past, we also will attain in days to come. Every one of us is a "divine fragment," every one of us an eternal Spirit, every one of us a deific life, striving to attain through matter to consciousness of our own divinity. That is the teaching of all faiths, that is the fundamental principle of life, of religion, of nature; and Modern Science is finding that even physical nature is not intelligible without the understanding of the higher world, without the recognition of larger possibilities.

Occultism, Semi-Occultism and Pseudo-Occultism.

A Lecture delivered on Thursday, June 30th, 1898, at the Blavatsky Lodge, London.

SPEAKING to the Lodge for the first time after returning from India, it will not seem to you, I think, either strange or inappropriate that I should take for my subject one which is largely drawn from Indian history; not the history of the outside nation, but the history of that inner line of thought which is of the deepest interest to us as students and as Theosophists. And inasmuch as history continually repeats itself, such a study may offer points of instruction to us in our own time. For I am going to ask you to consider with me what I may perhaps define – although definition is a little difficult – as, first, occultism, then what may be called semi-occultism, and, thirdly, the out-growths which follow and surround these and which are specially marked and active at any time when true occultism is working in the world.

It is a very common blunder made by many people to suppose that spiritual forces have in them something which they are pleased to call unpractical, and we continually notice an assumption, which is taken for granted without argument, that if a nation, for instance, should turn itself towards a spiritual ideal, or if individuals should devote themselves to the spiritual life, that then such a nation is likely to be undistinguished along other more evident and visible walks in life, and such an individual is likely to lose much of what is called his practical value in the world. Such a view of life is a blunder, and a blunder of the most complete kind. The liberation of spiritual forces, the setting free of energies on the spiritual plane, has a far greater effect both on the individual and on the nation in the other regions of its activity than can be produced by any of the forces that are started on the lower planes of life. When a spiritual energy is set free it works down through the other planes of being, giving rise on each plane to a liberation of energy, and bringing about results great in proportion to the nature of the spiritual force. So that it is true in history, as you may find by study, that when spiritual forces are liberated the intellectual life of the nation will also leap forward with tremendous energy, the emotional life of the nation will show fresh development, and even on the lowest plane of all, the physical, results will be brought about entirely beyond anything that could have been achieved by the energies of the physical plane which are set to work and which apparently cause these results. That is a principle, a law, which I will ask you to bear in mind through all that I have to say to you – that every force initiated on the higher planes, as it passes down to the lower, brings about results proportionate to itself; so that it is the shortest-sighted view of human life and of human activity which imagines that devotion to the spiritual life, the evolution of the individual in the spiritual world, is anything but an immense addition to all the forces of progress that work on the earth, anything but a lifting up of the world on the great ladder up which it is climbing.

But there is another principle that we must also bear in mind in our study, and it is this: that as forces are liberated on any plane, the results brought about by those forces will vary in their character according to those who utilize the energies after their liberation. As we have often pointed out to you here, energies on the different planes of nature are not what we call good or bad in themselves. Force is a force; energy is an energy. When we bring in the idea of good and evil, of right and wrong, of morality and immorality, these ideas are connected with the results brought about by individuals in the utilization of the forces. A time, then, of great spiritual energy, of great liberation of forces from the spiritual plane, will be marked to a great extent by activities of opposed characters on the lower planes of being, and those energies which are liberated on each plane may be taken up and used by individuals for what we should call either good or evil. The great mark of good or evil, looking at it from this standpoint, is the use that the individual is making of these forces, or such part of them as he is able to control; whether he is using them for the uplifting of humanity, whether he is regarding them as the divine energy which he may use to forward the divine purposes, or whether he is simply trying to grasp them for his own separate ends, striving to apply them to that which he desires to grasp and to hold, serving his own purposes without regard to the divine economy. This, then, as I said, we will bear in mind in following out, first, as a lesson, something of the past in India, and then in applying the lesson that thus we learn to the movement which we know amongst ourselves at the present day, that great spiritual movement which is manifesting itself in the world and of which the Theosophical Society is one of the potent expressions.

To begin with, what is occultism? The word is used and misused in the most extraordinary ways. H. P. Blavatsky once defined it as the study of mind in nature, meaning by the word mind, in that connection, the study of the Universal Mind, the Divine Mind, the study of the workings of God in the Universe, the study therefore of all the energies which, coming forth from the spiritual centre, work themselves out in the worlds around us. It is the study of the life side of the Universe, the side from which everything proceeds and from which everything is moulded, the looking through the illusory form to the reality which animates it: it is the study which underlies all phenomena: it is the ceasing to be wholly blinded by these appearances in which we so continually move and by which we are so continually deluded; it is the piercing through the veil of Mâyâ and perceiving the reality, the one Self, the one Life, the one Force, that which is in everything and all things in it. So that really occultism, in the true sense of the word, may be said to be identical with that vision which, as you know, is spoken of in the *Bhagavad Gîtâ*, where Shrî Krishna declares that "he who sees Me," that is who sees the One Self, "in everything and everything in me, verily he seeth." Such a study, if you understand at all what is implied in it, must necessarily mean the development in the one who sees of the highest spiritual faculties, for only by the Spirit can the Spirit be known. We speak continually of proving this, that, or the other spiritual thing. There is no real proof possible of Spirit, save through Spirit; there is no proof of the intellect, no proof of the emotion, no proof of the senses which is proof when you come to deal with the reality of the Spirit. Nothing of the nature of proof along those lines, whether sensuous, emotional, or intellectual, can be anything more than a suggestion, a reflection of the truth, an analogy which may lead us on the right path, but proof in the true sense of the word it never can be. And it has been written truly in one of the great Indian scriptures and repeated over and over again in the other scriptures of the world, that there is in the full sense of the word no proof of God save the belief in the Spirit, for only the Spirit that is akin to Him, and that is Himself, is able to know, is able to touch.

Now looking at real occultism as thus defined, realizing that no one can be in the full sense of the word an occultist save one in whom the spiritual nature is developed and active, we should, in our next step, be able to separate off from this true occultism very much that goes by its name both in the past and the present, amongst those who went before us and amongst ourselves to-day. But we should need, in separating off all these forms of so-called occultism, to distinguish between those which may be said, in a sense, to be stepping-stones to the real, which were intended as stepping-stones by those who gave them to the world and which may be used as stepping-stones and utilized for progress, and other forms which are not really included under the name of occultism in any true sense of the term, those things which H. P. Blavatsky once spoke of as occult arts, and which for many people seem to include everything they regard as occultism - arts in which certain forces of nature are utilized and in which faculties are developed on various planes in nature below the spiritual ; for there are worlds above what we call the physical, but still below the spiritual regions, with which the development of certain faculties brings man into touch, enabling him to control and utilize their forces. There are almost myriad arts and lines of study of this kind which ought never by any real student, by any one who is seeking the higher truth, to be included in his thought when that thought is turned towards occultism. And some of you might clear up much confused thinking on this subject if you would refer to the writing of H. P. Blavatsky on "Occultism *versus* the Occult Arts," where she draws the dividing line extremely clearly and shows the position that these occult arts hold, and should be recognized as holding, when we are dealing with human evolution.

True occultism, then, is that to which at first I would ask you to turn your thoughts, and its pursuit implies, as I have said, the development of the spiritual nature. Now the moment we speak of the development of the spiritual nature we must at once recognize that for the larger number of us that development must necessarily lie in the future, but that we may begin to work towards it to-day; that it is of enormous import to our true progress that we should recognize it and work towards it, and not, by misunderstanding the nature of that development, waste our time, waste possibly many lives, by following blind alleys and mistaken roads. The development of the spiritual nature must succeed – and this is one of the most important points that we can realize – must succeed the purification of the lower parts of our nature. We must be pure emotionally and intellectually. we must have reached a certain stage, at least, of the elimination of the personality before anything that can rightly be called spiritual progress is within our reach. No amount of mere intellectual development - and I will come back to that point, for I do not wish in any way to depreciate that most necessary line of human growth - but no amount of mere intellectual development will of itself bring about the growth of the spiritual nature. With the fundamental reason for that I shall deal more fully in a future lecture, but I must say in passing that the development of the spiritual nature and of the intellectual nature are on one vital point in direct opposition. The principle that we call the intellect is the analyzing, the dividing, the separative principle. The very purpose of its evolution is the building up of the individual, its root lies in the Ahamkâra, or the "I"-making faculty, it is that which limits, which defines, which separates, which marks off the man from every other man, which makes what we may call that coating of selfishness which is absolutely necessary as one stage in evolution, which is one part of our growth in this world. It is a stage through which all humanity must pass, but which, regarded by itself, makes all those illusions which the Spirit transcends, and gives the touch of apparent reality to the separated self, the antagonistic self, the self that covets and grasps and holds and sets itself against all others. So that what we might call the very principle of illusion is represented by this intellectual faculty.

Necessary as its evolution is, none the less it is on this point in antagonism to spiritual evolution; for spiritual evolution means the recognition and the growth of the One Self into manifested activity, first within that sheath which has been formed by the intellect, and then by transcending it and bringing about that realized unity which is the object of our human evolution. It is for this that we place the unity of mankind in the spiritual regions, it is for this that we proclaim the brotherhood of man as a spiritual reality; for the Spirit is one, and it is only as that unity is recognized, consciously known – not simply intellectually seen, but consciously realized – it is only as that is done that the spiritual nature is in course of evolution.

Inasmuch as the intellect is separative and the Spirit unifying, inasmuch as the one gives rise to illusion while the other transcends it, as the one is the source both of individuality and of personality, whereas the other is the source of that oneship which we seek and shall realize – you will readily see how in the course of evolution these two parts of the nature cannot be regarded as causally related in the strict sense of the term, and we cannot say that by the evolution of the intellectual nature the spiritual nature will inevitably develop. On the contrary, we have to learn that we are not the intellect, but are to use the intellect as an instrument; that we are not the separated self, but the One Self living in all. That is the object of our evolution, that the goal of our pilgrimage; and therefore occultism, which means the study and the development of the spiritual nature, must transcend completely the intellectual evolution. It may even in many of its earlier stages find, and does find, its bitterest antagonist, its most dangerous enemy, in that very maker of illusions that you may remember we are warned against in *The Voice of the Silence*, that most spiritual book which so many of us have found as opening up the path for us to the spiritual life. Recognizing this, we shall naturally look forward to the spiritual evolution as a thing to be worked for rather than to be accomplished from the stage at which we are at present. We should also be prepared to realize the immense difficulty of such an achievement, to understand how much will have to be done with the character and with the nature, how tremendous are the demands that we shall have to meet, before anything which in the strict sense of the term can be called occultism will be at all within our reach.

In the history of the past, where true occultism was the life of the world, where that great fount of spiritual life flowed from the Beings in whom the spiritual nature was wholly developed, when the world was drawing its light and its life from such Beings, it was obviously not possible that their knowledge, their powers, their work could be largely shared in by undeveloped humanity or even by the comparatively advanced humanity that surrounded them. Still less was it possible that any great part of their teaching or any true comprehension of their work and their methods could be known to the people at large; and yet it was necessary that links should be made, that steps, as it were, should be created. The result of this necessity was that men who were advanced – although in them the spiritual nature was not yet wholly evolved – men of great powers, who stand out in history as giants in humanity, strove to make possible for the advancing ranks of mankind some understanding of the upward path that should be trodden, some realization of the methods that might be adopted whereby approach might be made to the spiritual regions.

These men, great as they were, were not, as I have said, men in whom the spiritual nature was wholly developed, supreme, complete. Their evolution in many cases – and I speak with all reverence of those so much greater than ourselves – may even be said to have gone along one line in excess of other lines of their growth; so that one man might have enormously developed intellectual power but less perfection perhaps of moral character; another might have made great advance in devotion and might not have developed so much of intellectual force; another might be keenly alive to the religious necessities of man and not so much interested in his philosophical evolution; another, again, might have turned his attention towards the development of certain sides of man's nature which would touch the physical regions of existence, and even to the forcing of faculties in man, which, when built up from below, would bring him into touch with parts of the astral or the lower mental world, and might force those faculties and the part of his nature to which they belong in advance alike of mental and moral evolution. Along these various lines you will readily see that individuals might have progressed, and that each man would be characterized in his thinking and in his endeavour to serve mankind, by his own qualities, the attributes which he had specially evolved. So that looking back into the ancient history of India we find great teachers, Rishis as they were called, of many different types, each giving to the nation some great gift from his thought or from his knowledge, intended to help the more advanced souls of that nation towards progress which should end in spiritual evolution. Hence, to take one line of growth, the great philosophical systems which we find in Indian thought, such a system for instance as the Vedânta. Regarded as an intellectual system of pure philosophy, it puts in a magnificent intellectual form a view of the Universe, of the One Self, of the One Life, and of its manifestations as illusory in the deepest philosophical sense, that serves as an intellectual training, as a step which men must take in learning something of the mysteries of the universe. This system, when studied apart from the Yoga that alone can make it practical, may be classed under the head of semi-occultism. It is a system true within its own realm, a system intended to help forward the progress of mankind, only capable of being grasped, of being followed, of being studied by souls already advanced in mentality: but none the less it is not the spiritual truth; it is only an intellectual presentment of one aspect of it, an intellectual showing forth of one side of it.

It is a thing that must always be remembered, that the Spirit can never be expressed in terms of the intellect, that the One can never be grasped in the terms of the many, and that any intellectual presentment of spiritual truth must necessarily be partial, must necessarily be imperfect, must be, as has often been said, a coloured glass through which the white light is seen; a ray is passed through the prism of the intellect which breaks up the white light of the Spirit, showing it in varied colours as these scattered beams, each one of which is imperfect in itself. One, then, of the great gifts to ancient India coming in this way as the result of true occultism, as the result of the mighty spiritual life, was the philosophy of the Vedânta and all those intellectual systems intended for the training of man, and giving, so far as the intellect could give it, a view of the spiritual reality. But remember the saving clause, "as far as the intellect could give it." The intellectual view is only a partial view; and such a view, however much it may help man to see intellectually something of the possibilities of the higher life, can never make him realize it in consciousness, or give the true knowledge which comes alone through the evolution of the spiritual nature itself.

Along another line of activity would come the many schools of Yoga. These schools, as you well know, were exceedingly various in their nature. Some of them were designed to develop the higher intellectual consciousness in man by means of concentration, by means of meditation, and thus to bring him into touch with the higher regions of his being; they were intended to lead him, stage by stage, to get free from the body, to pass consciously into higher worlds, so that his consciousness might function in those more extended realms of being. And we find many of the teachings of Yoga – you may read many of these systems at your leisure, those which come under the great classification Râja Yoga – carefully adapted to aid the growth of the mind, the evolution of the loftier mental faculties, the rising on to the higher intellectual planes, the passing into states of consciousness far beyond the reach of ordinary humanity. They are, again, a stepping-stone offered, but still coming under this heading that I have called semi-occultism. Other schools were founded which dealt with man in different fashion, which strove to force his faculties from below, to force the evolution and the training of the astral faculties, to bring him first into touch with the astral world, to make him familiar with a part of the phenomenal universe closely allied to the material. These have generally been classed as the schools of Hatha Yoga, and in them various methods were employed dealing with the lower vehicles of man. By these methods the body was trained, was to a great extent purified and rendered an obedient instrument. The power of the will was also enormously developed, the man was taught to be master of his lower nature and so to take what in very many cases was a real step upwards, although we cannot include it in any sense of the term under the heading of true occultism.

It must be remembered when dealing with all these schools, when looking at them and striving to learn alike their use and their abuse, that it is a great thing for a man to become master of his passions, it is a great thing to subject the animal nature, to be able to stand unshaken, no matter what temptations may assail the lower man. And very, very many of these schools, which it is often the fashion in the West to scoff at and despise, have yet in them this element, that they at least recognize that man's intellectual nature should be master of his sensuous nature and that he should learn complete control over the body, complete control over the passions. And even along many of the darker lines of evolution, even in the schools that tread the path which all those who would reach the highest should most carefully avoid, it is none the less true that the subjugation of the lower nature is most rigorously insisted upon. It is only the ignorant who suppose that those darker schools are all given over to sensuous practices. Many of the followers of those schools lead lives which, so far as that side of the nature is concerned, might be taken as examples by an enormous majority of the men of the western world.

Now the whole of these different schools rose and flourished in ancient India as the result of the great downpouring from the spiritual regions on to the lower planes, and naturally they were used both for selfish and for unselfish purposes. But in dealing with all those schools of Yoga which train the intellect and develop the higher forms of intellectual consciousness, it is well to remember that they are real stepping-stones to the higher, and that it is a necessary stage of our progress that we should practise concentration, that we should use meditation, that we should be accustomed to contemplate intellectually and emotionally the ideals which appeal to us by their grandeur and their nobility. Those are stages in our upward path, and stages that very many of us might well be utilizing now, with a view to the higher growth, the deeper wisdom of the future. Men took up these varying lines of evolution, stirred fundamentally by the prompting of the Divine Life within them, ever seeking to raise them and to help their upward growth ; stirred, so far as they themselves were conscious, by the natural and rightful desire for higher evolution, for further progress, for growth in life. For, as we have often seen when we have been studying progress, we cannot leap at a bound to the heights of the spiritual life ; we have to climb step by step, we have to utilize the higher thoughts in us for the subjugation of the lower, and then in turn to outgrow that higher when a greater height comes within our sight and within our reach. We have learnt, as we well know, in our studies, that we may constantly eliminate lower ambitions by nourishing a higher ambition, and that though that higher ambition be still attached to the personality, or even transcending the personality, be attached still to the individual, it is none the less a stepping-stone, it is one of the ways by which we climb. It is well continually to kill out our lower by our higher desires, though even those higher in their turn seem lower as we are rising above them and greater perfection comes slowly within our gaze. So that this longing for a higher life, this desire to develop, this yearning for progress, had, and have, their rightful place in evolution ; and it is out of the ranks of those who feel these, out of the ranks of those who use the methods which make progress possible, that are taken those who are capable of further evolution. They learn gradually to transcend the hope for individual progress, and learn that that also, in the fullest sense of the term, is illusory, inasmuch as the separated life is an illusion and cannot exist as separate in the higher regions. The true life is the life which is spent as part of the Divine Life, pouring itself out for others ; and no life is true, no life real, no life spiritual, save when the very idea of the separated life is entirely transcended, and all the thought of the being, all the energies of the life, are poured forth as part of the One Self and no distinction is recognized. Service is then the natural expression of the life, helping is that in which the true existence is felt. But ere it is possible that this ideal can be even intellectually realized, some progress. at least, must have been made in transcending what we recognize as the personality ; and it was in order to make that possible to every man immersed in illusion, as all men have been and are, that the various methods were suggested by those who would fain help their fellows forward, as steps on the upward path.

Others, seeing in the religious instinct in man – in that side of his nature allied to the emotions, in which devotion finds its root and the possibilities of its growth – seeing in that his easiest upward path, gave to the world the various forms of religion in all their variety of adaptation to human needs, thus making the path upwards suitable for those whose constitution attracted them chiefly in the direction of love and of service. Seeing, then, that all these methods of growth were most active at the time when the real life was working at the heart of things, it will not be difficult to understand how, as that life found fewer channels for its expression in the world, fewer who were ready to transcend their own limitations and to give themselves wholly as channels of the Divine Life, all these methods lost their vitality and a great part of their usefulness. And so we find, in looking around the India of to-day, that many of those things that were living are now dead, that many of the systems that were vital are now mere shells, forming subjects for intellectual controversy or for individual pride, but no longer stepping-stones to the higher life. Here and there, still, some gleam of the true life survives, some real use is being made of these stepping-stones upward; but so far as the great masses of the people are concerned, mere shells and forms remain—evidences of what existed in the past, evidences, may we dare to hope, of what may be in the future.

It is hardly worth while to remind you that while semi-occultism may serve as a stepping-stone to real occultism, pseudo-occultism is generally a distinct obstacle and hindrance. Under this heading may be classed all the "occult arts," in the study of which many promising beginners have lost their way and wasted their lives. Geomancy, palmistry, the use of the tarot, etc., all these things are well enough for those who want to tread the byways of nature and to gather knowledge of her obscurer workings. They may be harmless, interesting, even useful in a small way, *but they are not occultism and their professors are not occultists.* A little success in their pursuit – and success does not demand high qualities of either head or heart – is apt to breed the most absurd vanity and pretentiousness, as though this dalliance with the Apsâras of the kingdom of occultism converted a commonplace man into one of its rulers, a mage. A man may be past master of all these arts, and yet be further away from occultism than is a pure and selfless woman seeking only to love and to serve, or a generous, clean-souled man, devoted to the helping of his fellows. And if these arts be turned to selfish purposes, or if they nourish vanity, their professor may find himself approaching perilously near to the gateway of the left-hand path.

Looking for the application of this to our present movement, the lesson springs easily enough to our gaze. Again, in our own days, a great outpouring of real life has occurred. again an effort has been made by those who are the guardians, the reservoirs, of that life for our humanity, to pour out the true spiritual energies for the helping and the uplifting of man in every region of his being, the manifesting again of the possibility of the real life. This has been marked by certain definite statements made from time to time, by hints thrown out here and there by her who was the special messenger in our own day of this possibility opening up for our own race. And there is one passage in that paper to which I referred at the beginning, which gives us in a phrase the reality of life: we are told that when a man becomes a real occultist he becomes only a force for good in the world. Here is a sentence that people read without realizing at all its meaning, a sentence that comes in the middle of many other statements, and does not strike with its full force on the unprepared mind and heart. For many things may be said which are missed for want of receptivity, and many truths are proclaimed which remain dark and silent, save to those whose eyes are beginning to be opened to see, and whose ears are beginning to be opened to hear. And that statement, which really puts the occult life in a few words, is one that most readers pass by without realizing its significance. There is no true spiritual life, there is no real occultism, until the man at least recognizes that the goal of his living is to become a force for good, and that only, in the world. He is no longer to seek his own progress, no longer to seek his own life, no longer to seek his own development—no longer to ask aught that heaven or earth or any of the other worlds can give him for himself. There is only one thing left within him, the longing to be of service; only one thing the motive of his being, to be a channel for the great life of God, to enable that life to be scattered more effectively over the world of man, and over all worlds where that life exists.

When that is recognized, even afar off, when that ideal first dimly dawns upon the human heart – come it by way of intellectual apprehension of its sublimity, or by way of devotional recognition of its truth – then for the first time the spiritual life stirs within the man, the first germ of the spiritual nature begins to quicken into life. And so we begin to realize that if true occultism could be reached and understood by any of us, we should have to begin the preparation for it by working at character in the way that every religion has taught. How often do we hear it said amongst ourselves, "we know all these moral truths, there is nothing new in Theosophy when it simply reiterates the old morality. When we are told to be unselfish, to seek to help others forward, to eliminate the personality, to kill out our faults, it is all an old story that we have heard to weariness. We want something new, we want some fresh knowledge, some facts of the astral world, some strange things of the mental region – that is what we demand from Theosophy, that is what we are seeking, and we do not desire to have pressed upon us these ethical maxims, these continual repetitions, these old-world stories which every religion has made familiar, and which we can hear from any pulpit." And yet the truth of the matter is that along that path only the spiritual life has been and is possible for man ; that the Divine Teachers who gave the religions to the world with their perpetual insistence on morality, gave them knowing the spiritual life, and knowing that only along that line the real progress of man into unity with God was possible. And when it was again declared by the lips of the Christ that only he might gain his life who lost it, that those who would be perfect must sacrifice all that they had, when he again reiterated the ancient teaching that narrow was the path and straight was the gate-way, he was only repeating what all true occultists have taught as to the necessity of the training for the spiritual life.

As progress is made, all those methods of Yoga which tend to help forward the individual, which are followed in order to gain progress, practised in order to evolve faculties, and used in order that the individual may go faster forward himself - all these are dropped, and Yoga is regarded, not as the means of self-evolution, as we are accustomed to regard it here, but as the using of great forces for the lifting and the helping of humanity, with utter disregard for the going forward of him who is using them, with no thought of progress on the part of him who is wielding them for the helping of man. For in truth all control of higher forces, all utilization of these vast energies, ought to come only within the grasp of man when he has transcended the personality and has learnt to use them only for the helping of all. We readily admit this in the common things of life, and recognize the difference between learning the use of an instrument and the mere holding an instrument without knowing how to use it. A pen, for example, is one of the most useful of instruments, but its utility depends upon the brain and the heart behind it, upon the knowledge and the skill that wield it ; and a pen in the hands of a child is of no more use than any fragment of wood that the same child might pick up to use as a toy in its play. Very much the same is the grasping of the forces of the super-physical world by those who have not yet conquered the lower nature, eliminated personal desires and consecrated themselves wholly to the divine service. They are, truly, picking up an instrument which may be used for the highest and noblest ends ; they are, truly, placing their hands upon a tool which in hands that know how to use it may serve for the salvation of the race ; but unless the spiritual nature be developed, that tool fails in its highest purposes, that instrument fails in all its noblest possibilities. And it has this peculiarity, that whereas the pen that I used as symbol might be comparatively harmless in the hands of the child, the grasping of those forces by one in whom the personality is not eliminated may become a source of danger alike to himself and others, and may tend to retard the progress of the race instead of lifting it upwards. That is why some of us who have learnt but the mere alphabet of these great truths lay so much stress – stress to weariness, as I know some of you think when I am speaking to you – on the moral training which must precede all attempt at occult study. H. P. Blavatsky gave us the same lesson when she herself said that she had blundered, in teaching part of the alphabet of occult knowledge, without insisting upon that old precept that the moral growth must come before the occult training, and that the character must be purified, raised and spiritualized before any one should dare to lay his hand upon the latch of the occult gate-way. Hence it is that those qualifications that we have so often studied are made qualifications for imitation; hence it is that there has ever been the demand that only the pure should enter, that only the selfless should come in.

If I have spoken of the past to you to-night, if I have reminded you that amongst us to-day the very outburst of the new spiritual life will cause activity on all the lower planes, it is because I would bring the experience of the past to reinforce a lesson so often given from this platform, it is because I would warn you of the dangers that surround us on every side – dangers that some of us are beginning keenly to recognize, and to recognize just because they have to some extent struck us, and have therefore made progress the more difficult. So that it is our duty as Theosophists, as would-be students of the science of the soul, to be careful that in all things character precede any attempt at the gaining of power, that purity, selflessness, devotion, utter self-surrender, be found in us ere we touch the Ark of occultism – for without these any success is a defeat, without these any attempt is doomed to failure. And surely it is better for us to learn from the experience of the past than by the bitter suffering that grows out of the personal experience of to-day ; better to learn by the authority of the great Teachers who have proclaimed the lesson over and over again, than to have to learn it by the suffering that follows from grasping powers ere we are ready to use them, from plucking the fruit of knowledge ere it is ripe for our consumption, from striving to rule ere yet we have learnt to obey, and from endeavouring to snatch at the mighty forces of the spiritual realm until we have learnt that great lesson of the Spirit – that only by giving is the spirit shown, that only by utter abnegation is the true life realized. As the very life of God in manifestation is a life that gives everything and asks nothing back, so those who would reach unity with Him and realize what the spiritual life means, must learn to give and not to take, to help and not to hold, to pour out without seeking or looking for return. Only as we learn that do we become fit candidates for the higher knowledge, only as the heart is thus rendered absolutely pure may we dare to face the presence of the Master, hoping that when " He looks at the heart He may find no stain therein."

The Light and Dark Sides of Nature.

From "Lucifer." October and November, 1896.

EVERYTHING in this universe of differentiated matter has its two aspects, the light and the dark side, and these two attributes applied practically, lead the one to use, the other to abuse. Every man may become a botanist without apparent danger to his fellow-creatures; and many a chemist who has mastered the science of essences knows that every one of them can both heal and kill. Not an ingredient, not a poison, but can be used for both purposes – aye, from harmless wax to deadly prussic acid, from the saliva of an infant to that of the cobra di capella.

H. P. BLAVATSKY.

In one of the scriptures of our race it is pointed out that at the very beginning of the universe the pairs of opposites appeared. "The pairs of opposites" may be taken as a general name for the light and dark sides of Nature, and a word on this general meaning of the pairs of opposites and on what they imply in Nature may fitly be said in opening.

First, it is impossible to think at all without pairs of opposites; we can only think, that is, by and through duality. If there were but a single thing undifferentiated, always the same, always everywhere, no thinking could arise in that thing. There must be at least two – the thinker and the thing thought of, distinguishable from himself, before what we call "thought" can exist at all. Not only so, but in thinking we find ourselves continually distinguishing one thing from another, we perceive the presence of these opposites: light and not-light, dark and not-dark – put in the most general form, A and not-A. To recognize identity – A = A, and to perceive difference, A is not not-A – is the condition of thinking, the law of the mind. Without this no mind, no thought can be. It is because this fact is recognized that in philosophic religious books the phrase which strikes many western thinkers as not only strange but nihilistic is used: Brahman is "without mind." So long as only the One exists nothing that the incarnate intellect can call "thought" or "mind" can be present. There is something deeper than "thought," something which is the root of "thought"; but thought as known by the brain must always imply duality, for without this we are unable to perceive, perception depending on distinctions.

While this formal statement may be unfamiliar it must at once be seen to be accurate when it is understood. For the very moment anyone thinks of anything he distinguishes it from other things by its differences, and assigns it relations by its identities; he distinguishes it from everything which is not itself, and he recognizes in it identities with things previously perceived, things to which it is akin. We only know things as we separate them by differences from the things they are not, and classify them with the things they resemble.

The pair of opposites that we are now taking for our consideration is the fundamental pair of opposites, one therefore of vast importance. This pair has long been called "the light and dark sides of Nature." It is the primary pair of opposites arising from the One, the fundamental duality, known to all students as being the nature of the second or manifested Logos. This second Logos in Christian phrase is the "Word made flesh," and in philosophic phrase apart from any special religion. is spirit-matter. male-female, life-form, positive-negative, the two aspects of the One between which the whole universe revolves. "Father-mother spin a web," the web of the universe. In this Logos, the manifested Word, the manifested God, the two poles of existence appear, and between these poles the universe is builded. They exist always together; they are co-eternal. one cannot be without the other. They are never known separated in Nature. Without the one the other could not be, could not even be thought. Fundamentally the same in their essence, they differ only in their manifestation. The whole of evolution is the progress of these two side by side, and evolution consists in the differences of proportion between the two. One is more manifest and the other less manifest: one is predominant and the other subservient; always, however, together in whatever part of the universe we may be. In the highest spiritual region life is not alone, but there form is so subtle that it lends itself to the slightest change of the informing life. In the densest region of the universe life is present; but there form is predominant, is rigid, unplastic, and the life is concealed beneath the rigidity of the form. Life implies consciousness, and form is that in which consciousness becomes manifest, and necessarily implies limitation. The two best words for this fundamental duality are really life and form, sometimes called in eastern books name and form. For name has in it the moulding potency and shapes the form it inhabits, therefore has name always been secret and sacred, and all potency lies in the "word." If "name-form" has become restricted to the lower plane it is because occult knowledge has been lost. Truly is it written that "Life is concealed beneath name-form," and these are the manifested universe.

Now the light side, the side of spirit, life or the positive, is the constructive and motive side; the dark side is the side of matter, form, or the negative, and is always subject to destructive transmutation, for only by destruction of forms can a fuller manifestation of life be made. Light and dark in nature then are the constructive and the destructive forces, both of which are necessary for the evolution of the universe, equally necessary, strange perhaps as that at first may sound. The light and the dark are equally manifestations of the One.

The light and the dark are equally necessary for the manifestation of the One. For without the light there would be no construction, and therefore no universe, no manifestation; without the dark there would be no destruction and therefore no evolution. For as each form is constructed it becomes a mould in which the life is held; and there could be no evolution, no progress in the universe unless that form can be destroyed and give place to a form which is higher and nobler. Within that form the life has been accumulating experience which has caused internal growth and differentiation. The form which expressed the life ere that experience was gathered now cramps its further growth and hinders its further expansion. If the life is to evolve, the form that imprisons it must be broken, and a new form must be constructed which will express the new powers of the life. Life is continuous, while forms are transitory and are shaped to successive stages of the life. The form that prisons is broken to set the life free to enter the form that expresses it. That also will become a prison in its turn to be broken, and so on stage after stage. Thus all evolution depends on the presence of this destructive side of the One Divine Existence, breaking down every form, not for the mere purpose of destruction, but because death is only the dark side of birth, and there is no death in one region of the universe which looked at from another region is not birth. Death and birth in fact are only two correlative names, and they are used in relation to the standpoint of the speaker. The passing of a life out of the region in which it is, is death to its form in that region; but as it passes out of that region by death it appears in the next region by birth. Therefore birth and death are rightly called the wheel of existence – both equally necessary, both equally fundamental; construction and destruction continually succeeding each other, both stages in evolution, and stages which are equally necessary. The manifested Logos, call Him by what name men will, is spoken of in all religions as creator, the unmanifested as destroyer; sometimes He is styled the regenerator, a name which includes both – creation and destruction being thus seen as the two poles of the one life, and in all manifestation these two are present.

The next stage in our study is an understanding of the three great regions to which the general evolution of ordinary humanity is at present confined, and it is necessary that it should be clearly understood that the question of good and evil does not come into play with regard to these regions *in themselves.* I want to get rid of the idea which is lurking in many minds that "spiritual" means "good" and "material" means evil. Spirit and matter, life and form, are never separated, and in themselves are neither good nor evil.

But spiritual is a name often used to define a particular region in Nature where form is dominated by life, just as much as material is used to indicate another region in Nature where life is dominated by form. Neither life nor form, spirit nor matter is good or bad in itself; both these poles are present in every plane, in every entity, and the entity is good or bad according to the end to which its activity is directed. There is good and evil spirituality just as there is good and evil materiality. The words good and evil have nothing to do with the fundamental constituents and forces of Nature, and people are constantly getting into a confused condition of mind because they take "spiritual" as meaning good; and then try to deal with the "dark spiritual side" of Nature, finding themselves face to face with what they recognize as evil, and yet find existing in the "spiritual" region. The forces of any region are non-moral, though both constructive and destructive entities are good or evil as they use these forces for or against the Divine Will. We shall avoid confusion, if we consider the planes of Nature as they really exist, and then define each clearly so far as it concerns us.

The word spiritual being used so loosely is apt to be misleading ; the third and fourth planes (counting from above downwards) form a region beyond the reach of moral evil, and if these alone are termed spiritual, evil would be excluded from the conception of spirituality. But the word is often applied to the mânasic, or intellectual plane, by Theosophical writers, and as the " Brothers of the Shadow " function thereon, its forces can be turned to evil purposes and are often thus turned.

The two highest planes of the septenary do not concern us, as human evolution in this manvantara does not touch them. We have thus left five : the âtmic or nirvânic ; the buddhic or turîyic ; the mânasic or mental ; the kâmic or astral ; and the physical. The âtmic and buddhic planes will only be touched by ordinary humanity in future rounds, so that for practical purposes we are confined to the three lower planes, the mental, astral and physical. In these man spends each of his life-periods, repeating the cycle over and over again.

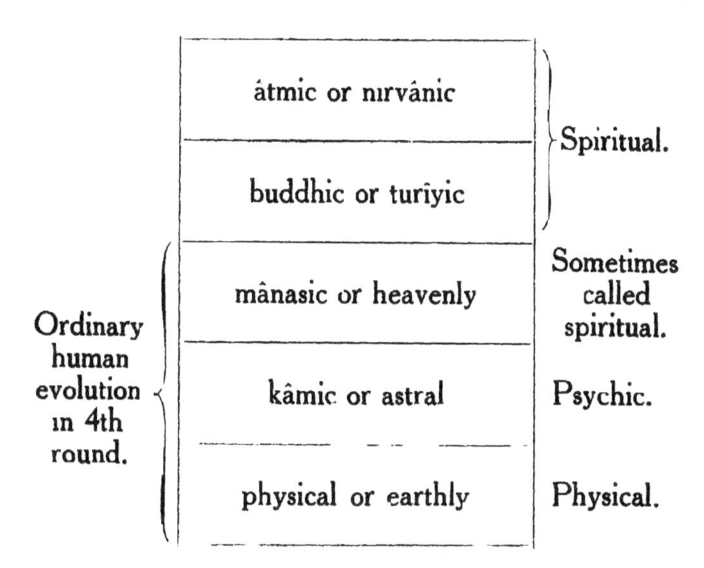

Thus for ordinary humanity we might name the three lower planes spiritual, psychic and physical, and in this way they are often distinguished, for they are the regions of heavenly astral and earthly life, and heavenly or devachanic life is that which satisfies all the part of man's nature usually regarded as spiritual. This use of the word spiritual brings us into line with the use of it by the different great religions, as with St. Paul's "spiritual wickedness in high places," the Hindu Asuras, the Buddhist Mâra and his hosts, the Occultist's Black Magicians or Brothers of the Shadow.

Further, the word spiritual is not inappropriate,

as in that region of the universe the spirit or life side is predominant, while the matter or form side is completely subordinate. Matter is very rare, very subtle, very ductile, very plastic, and it changes its form with almost inconceivable rapidity. Sometimes the higher region of this plane is even spoken of as formless. It is formless to everything which is below it, because the senses of the lower cannot appreciate the forms of the higher. But to those who are upon it form exists, for without form - which is fundamentally extension, that is, matter - manifested existence cannot be. The lower part of this plane is the region of the lower mind, but matter still remains quite subordinate to spirit, form to life.

In the next, or psychic plane, form is denser though still plastic, and life is more veiled. Both are active, but they are more balanced. Above, life is predominant; in the middle, life and form are balanced; in the lowest plane, or the physical, form is predominant and life is hidden. That is perhaps one of the simplest and clearest ways in which we may recognize the characteristics of these planes. In the first life or spirit predominates; in the second life and form are balanced - it is the battle ground of Nature: in the third, form triumphs.

In the lowest stages very many western people

do not recognize life at all; they regard it as one of the Theosophical follies to say that there is life in the lowest forms of material existence. But amongst some of our more advanced and younger chemists the phrase "evolution of metals" is being used, and "the life-history of metals" was lately spoken of in a lecture given at the Royal Institution. So that it looks as though science would soon no longer oppose the reasonable view, would begin to understand that life is everywhere in a universe which proceeds from life.

To pass now to good and evil. Everything which is in accord with the Divine Will – and in a moment I will define what I mean by that phrase – and which therefore works for progress and for happiness, is good; everything which works against progress and happiness is evil, no matter whether it be on the highest, on the middle, or on the lowest plane. It is not the forces which are good or evil in themselves, but the use that is made of them; not whether they are spiritual, psychical or physical, but whether spiritually they are used for good or evil, whether psychically they are used for good or evil, whether physically they are used for good or evil. The good or the evil depends on whether they work for progress or against it, whether they work towards the happiness and the perfecting of the universe or against it. On each plane there are forces which can be thus used. The forces in themselves are not good or evil, they are merely spiritual, psychical or physical. They become good or evil according to the purpose for which they are used, and the end which they bring about. Electricity, for instance, is neither moral nor immoral in itself. It is used immorally if it be employed to kill; it is used morally if it be turned to help and to comfort. And so in other regions of the universe. A spiritual force is evil if it be used against progress, for the causing of misery and of destruction; it is good if it be used for progress, for the bringing about of the happiness and the perfection of the universe. On each plane you may find good and evil, the distinction being in the use of the forces and not in the nature of the plane.

Let us now examine the meaning of the phrase "Divine Will." The one existence emanating a universe may be described as causing a great circle of existence, a vast cycle. It is said that "spirit descends into matter." That is a phrase consecrated by long usage and one to which there is no objection if its meaning be understood. But it is apt to be exceedingly misleading if people think of spirit as being somewhere up aloft, and matter separated from it somewhere down below, and spirit falling from above into matter. That is the conception which a good many people really have, though they might not put it quite so plainly. And as they think spirit climbs up again out of matter to where it was before, they not unnaturally ask, what is the use of the whole proceeding, why should it start if it is only going to return? Sometimes the Theosophist who is not quite instructed is apt to be a little irritated with his questioner, but the enquirer is really quite justified in his challenge, for where we Theosophists have expressed ourselves badly, it is quite right that a question put to us should point out the clumsy way in which we are saying what is yet fundamentally true.

There is a vast cycle of manifestation which may be regarded for convenience sake as a circle; at every point of the descent spirit and matter are side by side, but there is the change of proportion before mentioned, the spirit becoming more hidden and the matter more evident; the change in the ascending line is that the matter becomes more subtle and the spirit becomes more predominant. The Divine Will is the law of progress. This existence, manifesting itself, wills to bring a universe into existence, and to conduct that universe by evolution to perfection. It may of course be asked, why should it will to emanate? That is a question which we cannot answer fully, but we find in existence at the end of a universe a number of self-conscious individuals who were not in existence at its beginning, and who are capable of perfect life, perfect knowledge, perfect bliss. Even from our limited standpoint it must be admitted that this is a reasonable and sufficient purpose for the existence of the universe; it brings into conscious being these blissful all-knowing intelligences who share with the Divine Life that gave them birth its own existence, its own knowledge, its own joy. What a universe is to the manifesting life no words of limited mind may tell. What it gives to those who gain self-consciousness, bliss and knowledge by the process is sufficiently evident to any one who thinks at all. It is the difference between knowing nothing and knowing everything – a difference far more than between a stone and the highest archangel; for there is evolution behind the stone as well as in front of it, an evolution that prepares its existence as well as an evolution that carries it on into the highest ranges of self-conscious being.

Now this process at first and all through must be regarded as double – the light and the dark sides. One of the streams of divine energy is constructive, the other destructive ; one of them is life, building forms ; the other is death, breaking them up. Both are equally necessary, for destructive energy is going to destroy every form when it has served its purpose, in order that the materials used in the form whose purpose is over may be taken up by the constructive energy and built into a higher form. This process is what we call evolution. At every stage of the downward curve in which form becomes more prominent and life more veiled, forms will be brought into existence by this descending energy. Against it there will be working a destructive energy, which breaks up these forms as soon as their purpose is served, and they become outworn. There are thus two opposing streams of energy, by one of which forms come into existence and by the other of which they are constantly broken up, in order that higher forms may be built from their materials. There is no increase of matter it must be remembered, constant change, constant transmutation, but no increase and no diminution. Evolution consists on the form side of this process of destroying the lower forms that the higher forms may come into existence.

The next point is at first a little difficult to conceive, even a little startling. Growth is at first from the one to the many, from one existence to a universe of countless forms. "It willed, 'I will multiply.'" Then this descending line must be a process of separation, of making differences in order that an ever-increasing multiplicity of forms may be brought into existence. The key-note of evolution will be separation. As far down as the lowest or most outward point of evolution the key-note of progress is separation. The perfection of a universe is in the multiplicity of its forms, in the variety of the existences that are found in it. The universe exists in order to bring all these separated forms into manifestation, and all through this early process evolution will work for separation. Using the phrase the "Divine Will" - that will which is "I will multiply" - the Divine Will will work for separation, will work to make forms which are more and more separated and diverse from each other, in which the fundamental unity of life is more and more hidden. The whole of this growth will be a process of increasing separation. It is said to be a coming down into matter, and we may venture to use the phrase now that we have guarded ourselves against mistake; as things become more and more material they obviously become more and more separated. We may see that in the very density of matter as we know it. A piece of sulphur, for instance, is more separated from a piece of iodine than if both are sublimed to gas. The analogy is clumsy, for the gases remain separate molecularly though mingling in mass, so that there is no real union. But as we pass from the subtle to the dense this separateness of form is the thing that strikes us, whereas when we are dealing with very subtle things their unity is more prominently characteristic. If we understand this "descent into matter" we shall see that under the circumstances of the descending arc the opposition to progress would be the desire to remain one, would be the refusal to take form, would be the unit setting itself to maintain unity instead of accepting separateness. Hence setting itself against the Divine Will – wrong because it is against evolution, against progress, against the perfecting of the universe at this stage – would be, strange as it may sound, the refusal to take form in more and more material shapes.

Theosophists who have really studied may here see a gleam of light on what otherwise may have seemed to them strange; in the wonderful Stanzas of Dzyân it is said that the sin of the mindless is preceded by the refusal of the Sons of Mind to incarnate. That is, that the refusal of spirit, as we will call it for the moment, to take to itself separate form goes before the great sin which was wrought by the mindless men, and has left its traces in some of the higher animal forms. Intelligences awaiting incarnation set themselves against the law of progress. They looked on it as degradation to clothe themselves in the available bodies, as lowering their position to take forms in this lower world, and they refused to come down. Thence came the great primary transgression, known to students as "the sin of the mindless." To remain out of gross matter was against the law of progress, against evolution and the perfecting of the universe.

What at first seems so strange is that everything that now is right, the seeking of unity, the getting rid of separateness, the dominating of the material – at that stage of progress was wrong; the duty of these intelligences was to descend from the psychical to the physical, in order that a universe might come into existence in multiplicity of forms, in order that this building process might go on in which they were necessary helpers, co-workers with the Divine Will. Opposition to that Will, as ever, brought evil, but the nature of the opposition in this case was the refusal of spirit to enter physical forms, to veil its light in dense matter.

As the evolutionary process went on the spiritual was veiled in the psychical, and then the psychical was veiled in the material and the most material race of men appeared. Yet it was really a rising, descent though it seemed, for it was part of evolution, it was the way to the swifter bringing into existence of self-conscious individuals of our humanity. Without this the perfected manifestation would have been long delayed, without it self-conscious spiritual intelligences could not have developed so rapidly as the harvest of the universe, as the justification of this emanation of the Divine. Thus in this downward sweep of evolution what we now rightly call evil was then really good. To become separate, to become material was good in those far-off æons. For separation was necessary in order that a more perfect unity might finally be gained. Intellect could not evolve without spirit working through the lower forms of matter. The coming thus into the closest connection with matter of the physical plane brought into existence the human brain, the physical basis of all the faculties of the lower mind, and made possible the acquiring of the knowledge without which the individual could not expand into the divine.

In the process of evolution this lowest point was thus reached, and then there was a change. The utmost separation having been achieved, the utmost multiplicity of forms having been achieved, the utmost multiplicity of forms having been brought into being, then what we call the upward curve began. Life, having made this infinite variety of forms for its own manifestation began to work upon the forms to render them plastic. First the process of differentiation to get the forms, then the working in the forms to make them ductile as the expression of the life. These are the two great stages. The form must be brought into existence, and that means separation ; then there must be work from within to make the form the plastic expression of the life. The whole of the upward curve is used for that second half of the work. Life constantly toiling within these separated forms to make them more plastic, more transparent, working towards unity. Unity must be regained or immortality could not be achieved, for that which is composite cannot last for ever. But it is a unity into which has been absorbed the very essence of all the differences that have been passed through during the circle of evolution. The subtle life-form clothes itself in varied garments, subdividing and becoming more and more separate as it comes downwards, then a life-form separated from all other life-forms by this clothing of denser matter beginning the upward path in which it will work on its material garments, making them more transparent, more subtle, more a mere delicate film, and yet that film containing in itself the essence of every separated form through which it has passed. When at length it arrives on high, having passed into the intellectual sphere, it has in high and spiritualized forms the faculties which were latent in it at the beginning and has become self-conscious and not only conscious. Then it becomes one with others, but has the memory of its separateness behind it, reaching a stage which words must fail to describe, but which – borrowing a phrase from Madame Blavatsky – I may perhaps call "a conscious entity becoming consciousness." It keeps the memory which has made it an individual, and yet shakes off from itself everything which separates it from other individuals. It shares their experience and knows their knowledge, and yet is itself. It reaches the state which is spoken as of Nirvâna, which is the very antithesis of annihilation, which extinguishes separateness but keeps everything which by separateness has been gained ; it is the All, and yet in it is preserved the subtle essence of memory which was gained when each knew itself as one of many.

In this upward sweep, therefore, it is separateness which is to be gotten rid of, and therefore separateness is called "the great heresy," therefore it is called "the great sin," therefore it is the fundamental evil, therefore it has become the mark of what is called the Black Magician, the brother of the dark side. To keep the self separated from other selves, to seek everything for the separated self, and not for the common self of all, is now the worst sin. The Black Magician seeks for strength in order that he may be strong, whereas the White Magician seeks it in order that all may be strong. The Black seeks knowledge that he may be learned ; the White that all may be wise. When the White Brother reaches the spiritual plane, everything that he has gained in upward climbing becomes part of the general store, everything that he has gathered in his passing through the world becomes a common light which radiates in every direction. It is his own truly, but he has shaken off everything that separated him from others, he is able to shed all he has over the whole world of living things ; everything that he has gained as a separated self radiates out from him as an unseparate self to the universe of unmanifested existence. For where he stands there is no separation ; there is love, and love knows no separation : there is perfect wisdom, and perfect wisdom knows no separation. It is by ignorance that separation exists, and perfect wisdom clears away the veils that divide, and makes man realize that he has only become separate in order that he may gather, and has re-become one in order that he may give. In that region everything is common property. There is no longer "mine" and "thine," for all selves are one.

In the upward path then, the dark side will evolve by the desire to be separate, thus working for disintegration, and against progress. The Black Magician evolves by clinging to the separate form, by the desire to possess for the separate self. If that determination to be separate continues, if the desire to be apart from everything instead of being a part of everything persists as man rises upwards, then this one possibility remains : for a time by the tremendous strength that he has gathered, by the mighty knowledge that he has won, by the almost omniscience that he has gained in the long striving upwards, he can for a time, even in the spiritual region, hold his own against all others, for a time even in that world of unity can preserve a separated self. Not for ever, only for a while. He has won such tremendous force and energy and knowledge that he can hold his own for a time even against the Divine Will ; he can keep himself apart even against everything which tends towards unity. Even in the arûpa region of the mental plane there may be for a while separated existences which work for themselves, which are selfish, which refuse to hold for the common good and for the common enrichment, who are learned as separated selves, strong as separated selves, who use their strength to rule and to hold instead of to serve the world and lift it higher. Those are the great Black Magicians that are spoken of, the "Lords of the Dark Face," mighty in their power, mighty in their knowledge, mighty in the spiritual height that they have gained - very Gods in the manifested universe, but selfish Gods, anti-Gods, and therefore incapable of immortality. For only that can live which is one with the All, and they must break in time. The separated form built apart from its fellows and keeping itself separate whilst the universe is gradually becoming one, being against this upward trend, against the law of progress, against general evolution, is always striking itself against the law. It is deliberately dooming itself to disintegration, for the Divine Energy breaks up every form, and if it keeps separate form it also must be broken up. Though the dark spiritual powers – the God of the dark side of Nature, as they may have been called, the *Deus inversus* – may last for many an age, for many and many a millennium, yet as they have chosen form, and all form is perishable, they must at length perish. The forms that they have chosen must be disintegrated, and if they have identified themselves with the separated forms, then as forms they cease to exist. Having chosen the forms that perish, when those forms break, their consciousness goes back into the vast ocean of consciousness ; they have failed to extract the essence, to transmute it into consciousness *per se* ; they have chosen a self-conscious individuality which is separate, and when the separateness breaks, the consciousness goes back into the ocean and self-consciousness is lost.

Any, if he will, may choose that side. We all of us are choosing it from time to time. For every force that works for disintegration works for its own destruction ; every force that at this period of the world's progress works for its separate self is throwing itself against that mighty stream of destructive energy which breaks and grinds everything to powder in order that it may be rebuilt anew of higher mould. Every agency which works against the whole, everyone who separates himself and works for himself against his brothers, every such force is a force that is working for self-destruction, destruction which is self-chosen and which Nature cannot refuse to give.

Now we can realize what evil means. Evil is everything which works against the Divine Will in evolution. It is everything which works against truth which is God, against unity which is God, against love which is God. Every such force is working against the whole, and if it comes into conflict with the general force which is working upwards, and with those who are the embodiments of that upward tending force, it must inevitably be broken into pieces. The Great White Lodge wars against none, but it goes its way, and that which wars against it is broken into pieces. It does not war with hatred, it passes upward ; it does not use the weapon of wrath and of anger, but it passes upward. Everything which flings itself against it is, by its own act, and not by the act of the White Brotherhood, broken into shivers ; it breaks into fragments, while the great force goes on.

Some imagine that the force of the White Lodge is used for destruction, but it is not so. That Lodge is on the upward arc, and the White Brothers are ever on the side of unity ; where there is conflict it is the disunity flinging itself against the unity, and as that is unchanging and ever going towards its end, those which fall against it are broken into pieces. Here is the occult meaning of a phrase which is familiar in the Christian Scriptures, that those who fall upon the stone which is the head of the corner are broken ; not by the action of that mighty corner stone, but by their own action ; not by its disruptive energy – for of disruptive energy it has none – but by virtue that it is changeless and cannot be broken, and that everything that works against it must shiver from the energy with which it flings itself against the law. The whole mighty sweep is the law which passes downwards and then upwards once again. Everything which is against it is broken, everything which separates itself from it must fall to pieces. Every separated existence must break ;

only in unity can life proceed, therefore when we study the light and dark sides of Nature in their bearing on our practical life we find that every force of hatred, of disruption that makes against unity, that works for separated fragments and not for one mighty whole – everything that works on that side is under the Black Lodge, is an agency of the Black Brotherhood. When we speak of the dark side of Nature and of those who incarnate the disruptive forces, as the White Brotherhood incarnates the law, the good law of the universe, we know that everyone of us must be on the one side or the other – working for brotherhood or working against it, working for construction or destruction, for building or for breaking, for unity or disuniting.

That is the practical outcome of this study: each of us in striving to lead a life which we would fain should lead us on the upward course and bring us at length into that unity wherefore the universe exists, will do well to scrutinise our own hearts and our own lives to see whether the forces in us are tending to Truth, to Love, to Unity. Everything that is of these is white. Everything that is against these is black. We *must* co-operate with the one side or with the other, and according to our final co-operation will be the final end of the individual soul.

The Destinies of Nations.

From "The Theosophical Review," Volume XXXVII.

CERTAIN great ideas, necessary for the evolution of the race, may be said to belong especially to the civilisations of the East, and those ideas were in danger of being trampled out by the advancing western civilisations. That was a danger to humanity at large, the ideals of both eastern and western civilisations being necessary in the future of the world ; and it became necessary for some definite interference to take place to re-establish the balance of thought. I want to draw attention to the nature of that interference, to show what lies behind the destinies of nations, and what forces guide the current of affairs, so that we may see through the veil of events to the forces that guide them. The great world-drama is not written by the pen of chance, but by the thought of the LOGOS, guiding His world along the road of evolution.

In the course of that evolution many beings are concerned. We have to look on this world as part of a chain of worlds all closely interlinked, all the inhabitants of these different worlds having something to say in those parts of the drama which are being worked out in each. We are all living in three different worlds, and not only in one; and whether in the physical world, or in the next world, the astral, or in the third, the heaven world, the inhabitants are busy with the general conduct of affairs which affect all three. Life becomes enormously more interesting when we recognise that it is shaped not only in the physical world but in other worlds as well, and that when we trace the destinies of nations we find that those destinies stretch backward, and that the working out in the present is largely conditioned by the energies of the past.

Let us look for a moment on the rough plan of the whole. Let me put it as though it were a great drama written by a divine pen. The story of the world, and the various parts of the actors on the stage, are all therein written. What is not laid down is who the actors shall be, and with regard to this a large amount of what is called choice comes in. This drama is the manifestation of certain great ideas in the

Divine Mind, ideas written, as it were, in the heavens; for it is suggested in very ancient thought that what we call the signs of the zodiac have a definite connection with the course of human affairs. Of that, in the broad outline, there is no doubt in the minds of any who have penetrated somewhat behind the veil. The importance of these starry influences cannot be over-estimated; for inasmuch as human beings are related in the composition of their physical and other subtler bodies to the worlds among which they move in space, there must be magnetic relations existing between them and the system of which they form a part, and at certain epochs in the history of evolution there will be one or another dominating influence present in the atmosphere in which men think and act, and they can no more escape that influence than their bodies can escape the influence of the far-off sun. The great drama, then, is the grand plan of human evolution. It is full of parts which are to be played by the nations, but not necessarily by this or that nation; for the nation qualifies itself to play a certain part which may be offered to more than one nation, and one or another may rise to the height of its great opportunity.

Leaving that for a moment, let us ask a

question as to the forces which help to adapt players to parts. Are there to be found, in what seems the great chaos of human wills, any guiding forces which bring the actor and the part together? You cannot well have a drama vast as the world-process, as evolution, and then a great gap between the Author of so vast a plan and the individual players who make up the nations and choose the parts. How is the right player to be brought into touch with his part in the history of the nation, in the history of individual successive births aud deaths? That is the next point to grasp.

Now the vast machinery for bringing together the parts and the players is found in the hierarchies of superhuman Intelligences recognised in all the religions of the world, and in the occult teaching on which they are founded. Not one great religion of the past or of the present that does not see surrounding the world and mingling in its affairs the vast hierarchies of spiritual Intelligences into whose hands is put the work of bringing together the players and the parts. You will see, if you turn to the religions of the nations of the past, how they have recognised these workings as playing a great part in the practical shaping of the destinies of nations. Not one great people of antiquity that did not have its own national "Gods."

The word "Gods," however, as used in the English tongue, is very confusing, for it is applied not only to those great hosts of Intelligences, but also to the Supreme, the LOGOS, the Author of the drama. Now in the nations that have other religions than the Christian this confusion does not arise. It is when the Christian is contemplating those whom he calls the "heathen" that the greatest confusion arises, for over the whole of their vast theology he uses the one name "God." And yet he might easily escape that by remembering that his own cosmogony is only a reproduction of the older thoughts of these more ancient peoples. In the East there is one name which is used for these Intelligences – the name "Devas," from the root "*div*," to "shine" or to "play"; the "shining ones," or the "playing ones," would be the English translation. When Bunyan so often used the term "shining ones" he was using a quite eastern phrase, for it is by that name that the East knows this great hierarchy of Intelligences. Among the Christians and Mussulmâns, whose religions are drawn largely from the Jewish, the name "Angel" is used, the terms "Angel," "Archangel," "Cherubim,"

"Seraphim," and so on, being represented in the older faiths either by the word "Deva" or by a word derived therefrom. "God," in the Christian sense, is known by other names, and no confusion arises.

In all the old religions these Devas played an enormous part, and each nation had its own particular set of Devas. The Egyptians regarded certain superhuman Intelligences as their earliest law-givers, and the connection between the human law-giver, the Divine King, and the Deva is always clearly marked. Every civilisation takes its rise in a little group, partly human, partly superhuman, to which it looks back and from which it draws its laws. The Greek had his Demigods or Heroes, and his Gods or Devas. So among the Chinese, the Japanese, the Persians, the Indians, the same idea is found of the nation being founded by the group which contained the human law-giver and the Deva who worked with him in the building of the nation. Celsus hints that the Beings "to whom was allotted the office of superintending the country which was being legislated for, enacted the laws of each land in co-operation with its legislators. He appears then to indicate that both the country of the Jews, and the nation which inhabits it, are superintended by one or more beings . . . . co-operated with Moses, and enacted the laws of the Jews" (Origen. *Con. Cel.* V. xxv.).

Now the Divine Kings, the Heroes, passed, but the Deva remains still at the head of each nation, a real existence in the astral and heavenly worlds, with a crowd of less developed Intelligences under his guiding hand. And when you come to the Jews you find that idea very clearly laid down in their scriptures. I pause for a moment upon it, because the sentence I am going to take from the Old Testament, from Deuteronomy, gives exactly the idea which I want us to take in considering the working out of a nation's destinies: "When the Most High divided the nations, when He dispersed the sons of Adam, He set the bounds of the people according to the number of the angels of God; and the Lord's portion was his people Jacob" (Deut. xxxii. 8, 9, *Septuagint*). To many modern readers the latter part of that sentence, "the Lord," may sound surprising, for they are accustomed to connect that word with the Supreme God; but we can see from the whole of the sentence that it is the name "Most High" which indicates the LOGOS, the manifested God, and He divides all the nations of the world according to the number of the angels, and to one great angel, "the Lord," He gives Jacob, Israel, as his peculiar portion. Origen, in dealing with this, alludes to the "reasons relating to the arrangement of terrestrial affairs," and points out that in Grecian history "certain of those considered to be Gods are introduced as having contended with each other about the possession of Attica; while in the writings of the Greek poets also some who are called Gods are represented as acknowledging that certain places here are preferred by them before others" (*Con. Cel.* V. xxix.). And so he points out that after what he regards as the symbolical dispersion, at the building of the tower of Babel, the different nations were given to these groups of celestial beings (*Ibid.* xxx.). These beings were worshipped in their respective nations, who followed their own "Gods," and not those of other peoples (*Ibid.* xxxiv.).

This idea of "the ministry of angels" is very general among the early Christians; thus we have in *Hermas* the vision of the building of a tower:

"And I answering said unto her, These things are very admirable; but, lady, who are those six young men that build?

"They are, said she, the angels of God, which were first appointed, and to whom the

Lord has delivered all his creatures, to frame and build them up, and to rule over them. For by these the building of the tower shall be finished.

"And who are the rest who bring them stones?

"They also are the holy angels of the Lord; but the other are more excellent than these. Wherefore when the whole building of the tower shall be finished, they shall all feast together beside the tower, and shall glorify God, because the structure of the tower is finished" (*1st Book of Hermas*, Vision iii., 43–46).

Clement (*1st Epistle*, xiii. 7) quotes the text above referred to. Also the following remark about Jesus, made by Satan to the Prince of Hell, is noteworthy: "As for me, I tempted him, and stirred up my old people the Jews with zeal and anger against him" (*Gospel of Nicodemus*, xv. 9). The Jews were under Saturn, or Jehovah, according to Origen. The same idea is taught among the Mussulmâns. They regard the angels as taking a very active part in the affairs of men. And it is hardly necessary to remind you that in the great epic poems of India, the *Mahâbhârata* and the *Râmâyana*, you find the Devas mingling with the affairs of men, so that when great quarrels are to be decided they manifestly take part in the strife, each struggling for the particular tribe or nation placed in his hands for its evolution. A correspondent, Mr. Tudor Pole, of Bristol, tells me that there is an old Teutonic legend that on New Year's Eve all the "Inner Rulers," the angels, of the nations assemble before the Council of the Gods to receive their orders for the coming year; each has his request to make as to the destiny of his nation during the coming year; the Council arranges the part that each nation shall play during the ensuing year, and the Great Lords are consulted. Finally, the Rulers disperse, some with music and joy, some weeping, some in great agony.

In Greece there is much mingling of "Gods" and men, and the Greeks, despite their philosophy, took the matter as real, not as fairy-tale, although the philosophers in Greece, as among the Hindus and Buddhists, did not worship these "Gods." In the 7th book of the *Odyssey* we read how "Minerva meets Ulysses, in the likeness of a young maiden bearing a pitcher," and she guides him to the palace of Alcinous, a palace guarded, in Atlantean fashion, by immortal gold and silver dogs, made by the mind of Vulcan. And so again in many another tale. written when men's minds were less blinded than they are to-day.

Of course, in modern times this idea has disappeared, and it must seem like a fairy tale to modern readers when one brings such thoughts into touch with what may seem to them such much more real things, the strifes of Kings, and the politics of the modern world. And yet behind all these the co-ordinating forces are still continually at work ; and when the time comes for a nation to play a triumphant part in the current history of the world, then, many years before the time of the triumph, there are guided into that nation by the Devas souls which are fitted for its building up and guidance in the coming struggle. And when the time comes for a nation to sink low in the current history of the world, there are guided to incarnation there souls that are weak, undeveloped, cruel, tyrannical, having fitted themselves to fill such actors' parts in the great national drama. Let us keep, then, that theory in mind : the drama on the one side ; this great co-ordinating agency on the other, guiding the self-chosen actors to their appointed parts.

And now let us look at some of the nations themselves, and see how far the destinies that they are working out fit in with this view of a guiding hand behind the veil. Let us take for one instance the building up of a mighty western empire, so that the great Fifth Race, with its evolution of the concrete mind, might play its part in the drama for the benefit of humanity at large. And let us see, if we can, whether certain definite currents may not be traced which show a plan definitely worked out, and not the mere mingling of the chaotic wills, ambitions, and selfishnesses of nations.

Slowly was prepared this part of a nation to stand high above the nations of the world. The first nation to whom that part was offered was Spain, who had been preparing for it by a very marked and extraordinary evolution. Into her was poured the great flood of learning which linked itself with the dying philosophy of Greece, and drew its rich stores from the Neo-Platonic schools; into Southern Spain came the great incursion from Arabia, rich with all the knowledge brought from the mighty schools of Bagdad, which spread over Southern Spain and thence over Europe. To her was sent Columbus, who made it possible for her to spread her conquering troops across the Atlantic and subject the new world to her imperial sceptre. How did Spain meet that wondrous opportunity? In the wake of

Columbus came the army, subjecting Mexico and Peru to her sway, and destroying their ancient civilisations, outworn and ready for destruction. She had laid upon her shoulders the task of building up in that new world a civilisation based on the solid foundation left there by Atlantis, capable of supporting the structure of the new thought and knowledge. All know how she missed her opportunity, how she drove out from her own country the Moors and the Jews, the inheritors of the knowledge, the philosophy, and the science; and how, in the new world, with her greed of gold, she cared nothing for the peoples placed in her hands, but trampled them into the dust. So her part in the drama was taken away and offered to another people.

Another nation became a candidate – a nation which, with many faults, had also many great virtues. England, spreading abroad her race, more and more subjected to her sway land after land. She gained the offer of a world-empire by an act of national righteousness – the liberation of the slaves from bondage, accompanied by that great act of national justice which sacrificed no one class but placed the burden of the liberation on the whole nation. For that, those who guided her destinies were offered the possibility of world dominion. All the nations that tried to establish themselves in that great land of the East, India, one after another failed, until the English race placed its feet therein. The story of the placing is not good to read, and many crimes were wrought, yet on the whole the nation tried to do its best and to correct the oppressions wrought in India – then so out of reach - as witness her action towards her great pro-consul, Warren Hastings, when she brought him to trial for his evil deeds, in the face of the world. So, despite many faults, she was allowed to climb higher and higher in the eastern world, partly also because she offered, with her growing colonies and language, the most effective world-instrument for spreading the thought of the East over the civilisations of the West. All know how far that has gone, how all over Northern America, in far-off Australasia, as well as in her own land, eastern thought and philosophy have everywhere penetrated, so that the treasures of Sanskrit learning, kept so jealously until the time was ripe for their dispersion, are being spread over the surface of the globe.

Continually, by lessons ever repeated, those Higher Ones who guide the nation are striving to impress upon England the lesson that by righteousness alone can a nation be exalted in the long run. And in a critical moment, when luxury was growing too enervating, too selfish, the terrible lesson of South Africa branded on the English conscience the lesson that duty and right must go before luxury. Through the fires of disaster a lesson was taught to England which, may God grant, she has learned for her future guidance.

And then there came the question of what nation should be chosen for the work of lifting up those ideals of the East of which I wrote last month. India, at this stage of the world's history, could not do the necessary service: she was learning her lessons under a conqueror; but there was a nation in the Far East which had within it the possibility of learning the lesson, and the Devas of the nation began to concern themselves with the attempt to train up in that far-off island a people who should be fit for the mighty task of uplifting eastern thought, of showing that conquest might go hand-in-hand with gentleness and self-control, and that a nation might spring into a mighty power without losing its sense of duty. The work began by a change in the education of the people, which might make a nation conscious of itself, and then into the soil thus prepared a group of heroic souls was born.

The Mikado of Japan, a mighty soul, fit to incarnate for that nation its own greatness, fit to use such power that in brief space of years he might transform the nation, put it into new shape, evolved it in unknown forces, and at the same time showed out a personality so wonderful that all that nation look to him as ruler by Divine Right, from whose sacred person flow the powers which in the nation are shown forth, every triumph reflecting new glory on his personality. And round him gathers one great one after another, for the labour of raising up the nation, until at every point of importance you see a statesman, a general, an admiral, fit to lead a people from triumph to triumph. A group of strong souls is guided to incarnate there, in order that the nation may fulfil its destiny ; for no nation can be great unless at the centre there be an ideal, and a perfect loyalty and self-devotion. It is no mere lip phrase, but voices a feeling deep in the heart of the soldier and of the general, when they thank their Ruler for the victory in the field, and with the eastern devotion say that he is the representative of God amongst them.

Glance at the other nation in the great duel which is being fought in Eastern Asia, and see how strangely Russia, a nation with a great future before her, is being guided through the frightful valley of humiliation. The preparation for that calamitous part in the drama lies in that which has gone before, even within the limits of our own lives. There was a moment, some twenty-five or thirty years since, when a wondrous opportunity came in Russia's way. Although ill-judged, there was a noble impulse underneath the freeing of the serfs, and there was a possibility that that act might be turned to good purpose for the nation, and raise it higher, instead of leading it wellnigh to destruction as it has done. And then there came out of many souls born just then among the nobles of Russia, one of the most wonderful things the world has seen – a flinging of themselves out of their own rank down amongst the poor, the ignorant and the down-trodden, a giving of themselves by the lads and the girls of the nobility to the lifting up of the people, not by a far-off charity, but by a wondrous impulse of uttermost self-sacrifice. And how was that met? The divine compassion of those youths and girls was met by the fortress of Peter and Paul, by the mines, and deserts, and snows of Siberia. Nothing more terrible has been wrought by a government of any people within modern times. And terrible the Nemesis.

Driven by despair, their attempts to uplift in all gentleness met with the knout and the underground dungeon, with starvation for the men, with dishonour for the women, what wonder some of them went mad ! What wonder that some of them at last, after years of patience, after cruellest sufferings, answered with the bomb to the knout ! This state of affairs was created in the first place by the bureaucracy and not by the victims. Thousands upon thousands of those who would have redeemed Russia died on the scaffolds, were slaughtered in those frightful mines, until at last the patience of the Gods grew exhausted, and the time came for the government to learn that governments exist for the helping and not for the crushing of their peoples.

So that Russia chose by her past that terrible *rôle* which now she is playing on the stage of the world. Against her are all the forces that make for progress ; against her from the astral world the myriads that she sent there before their time – all her martyrs, all her victims, are struggling against her. Hence the record of unexampled defeat. And at home, revolution, anarchy, assassination and mutiny are threatening her government fabric from every side, until for Russia at the moment there is only [illegible]

Valley of the Shadow of Death to be trodden from end to end ; and with pain at heart, but with steady hands, her angelic guardians guide her through the defeat and the disaster, willing that their charge should learn her lessons whatever the price she pays. For in those clearer eyes the nation's agony for the moment matters little, beside the lessons that through that agony are learned ; and until the tyranny itself is crushed, and the rulers of Russia learn their duties to the people, she must still tread the winepress of the divine wrath.

And see how Russia has been prepared for it. Among all her rulers not one strong man ; weakness and uncertainty everywhere, changed policy at every moment. Mark the government of him who should be the father, but is the tyrant, of his people – perhaps not a bad man in himself, but utterly unfit for his post. It is part of the destiny of a nation that, when the hour of its doom strikes, nothing but weakness is born into its governing classes, so that those who would not rule aright may lose the power to rule. And on those terrible battlefields of which we have read records in the daily press, is there anything more pathetic than the dauntless courage of the soldiers, and the hopeless incompetence of the officers ? It is not that the soldiers do not fight, but that they are led by men who know not how to lead.

It is thus that nations are guided from above, and into the nation that has to go downwards those are guided who inevitably drag it downwards. The same was the case in Spain – a child King, and not one able man among the ministers who could guide it right in the struggle with Cuba and America.

And how are these leaders chosen? They are chosen by their own lives in the past. A man is found unselfish, brave, and noble, and such a one, in the countless choices of his daily life, is making the choice for the splendid part that hereafter in humanity he shall play. And so with those who are great outside, but have to play a sordid part. By countless selfishness and preferring of themselves, by taking ever the lower path instead of the higher, those men choose also their parts in history.

Thus it is that the Occultist looks on human history, and sees preparing around him on every side the men and women who are to be the players of the future in the more prominent parts of the world-drama. For none forces upon us any part, nor imposes upon us any special place in the world-drama. We choose for ourselves. We build up ourselves for glory or for shame, and as we build so hereafter shall we inevitably be. Hence it follows that for a nation to be great its citizens must slowly build up greatness in themselves. Hence it is that the greatness that you see now in Japan is a greatness that you can recognize among her ordinary men and women, who are willing to sacrifice all that is dearest for the sake of their country and the glory of their chief.

And so with England, if she would fill the mighty part which is before her in the near future. She must build up her sons and daughters on heroic models, by placing righteousness above luxury, thought above enjoyment ; by choosing the strenuous, the heroic, the self-sacrificing in *daily life*, and not petty enjoyments, small luxuries, and miserable sensual gratifications. Out of rotten bricks no great building can be built, and out of poor material no mighty nation may be shaped. The destinies of nations lie in the homes of which the nations are composed, and noble men, women and children have in them the promise of the future national greatness. And as we make our conditions better, higher and more evolved souls shall be born amongst us. While we have slums and miserable places we are making habitations for little evolved souls, whom we draw into the nation. Under the ground the root grows, out of which the flower and the fruit will come, and poor the gardening science which places a rotten root in the ground and expects from it a perfect flower and a splendid fruit. If we would have England great among the nations, and make her destiny an imperial destiny as the servant of humanity at large, we must cultivate the soil of character, plant the sound roots of noble, righteous, simple living, and then the destiny is inevitable, and the nation will be cast for an imperial part in the drama of the world.

The Hathâ-Yoga and Râja-Yoga of India.

From "The Annals of Psychical Science," November, 1906.

IN the first place, allow me to explain why I have chosen this subject for discussion: I have lived in India for twelve years; I have made a fairly thorough study of Indian psychology. I thought it might be useful to speak about those matters of which I have some knowledge, and which are but little studied by the Western world.

There exists in India a psychological science, the origin of which dates back thousands of years.

It is known that India possesses a very ancient literature. Now, everywhere in that literature we find traces of psychology, and also the exposition of an ancient psychology, in its practical, and not merely in its theoretical aspect.

Since this science has been put into practice

for so long a period, is it not reasonable to conclude that there may be in these ideas, these theories, based on repeated experiments, something which may prove useful to modern psychology ?

This psychological science of the East is called Yoga, a word signifying to bind, to unite. When we speak of Yoga, we express the idea of forming a union, of binding. Of binding what ? Consciousness itself, by realizing the union of the separate consciousnesses of men with the universal consciousness. Yoga includes all the practical methods by which this union may be attained.

Yoga is thus a science which may be both studied and practised ; it is practised in order to obtain a complete union between the ordinary individual consciousness of man and the superconsciousness, by rising from plane to plane, until at last this union is completely attained : then one is said to be free.

In order to understand this science, and also the experiments which I wish to explain, allow me to give a short account of the fundamental ideas on which these experiments are based. It is probable that you will not accept these ideas ; but you may, nevertheless, understand them as theories : theory concerning man and, more particularly, theory concerning the consciousness of man. The theory, then, must be considered first of all, in order to be able to explain the aim; otherwise the experiments of the East will always remain unintelligible to western minds. If you will accept these theories, for the moment, you will understand the *ensemble* of these experiments, and you may perhaps deduce for yourselves conclusions from them which will afford clues with regard to other experiments. Herein lies the value, for western minds – so it seems to me – of a knowledge of this science of the East.

The first proposition is, that consciousness is one and universal. Everywhere, beneath appearances, behind phenomena, a consciousness is revealed; under the diversity of forms persists the unity of consciousness; a unique energy, a unique force, is everywhere in the universe.

This theory may be regarded as closely related to the western conception of one single energy of which all the forces are but the manifestations, the example. But, in India, this energy is always regarded as conscious, that is to say, no division is made between consciousness, life and energy; these are but three words denoting the same essence, but which establish also a distinction between the manifestations of this essence, a distinction which it is useful to remember when experiments are being made. But it must be recognized that this energy is one, and is conscious; is, in fact, consciousness itself.

The second proposition is that this energy, this consciousness – I prefer the word consciousness – manifests itself in the universe through the different forms of matter. The manifestations of consciousness depend on those forms by which it is conditioned. The differences which are perceived are simply differences of form and not differences of consciousness. Consciousness is always present, but it cannot express itself in a complete manner in a restricted form. The evolution of forms depends on this manifestation of consciousness; and when we place side by side consciousness and form, energy and matter, it is consciousness which directs, which is sovereign, which disposes of matter, and each functioning of consciousness creates a form for its revelation. When I use the word "creates," I do not mean creation out of nothing; I mean that consciousness disposes of matter so as to express itself, that all the powers reside in consciousness, but that in order to reveal, to manifest its powers, it it absolutely necessary to find the vehicle of consciousness, that is to say, to organize the material by which it can express itself.

I may on this point quote a very ancient line of a Upanishad, the Chhândogya : "The Self, that is to say consciousness, desires to see : the eye appeared ; it desired to hear : the ear made its appearance ; it desired to think : intelligence was there" ; that is to say the efforts of consciousness are shown in obedient matter, directed by that energy which incarnates itself in forms.

You will find the same idea in the physical universe, in the transformations of electricity. You may make different instruments to enable the energy called electricity to manifest, the energy is ever the same, it is only the manifestation that varies. According to the instrument you provide, you can obtain light, sound, heat, chemical dissociations, all these being merely manifestations of electricity, manifestations which are possible because you have provided instruments which afford suitable conditions for each kind of manifestation. But the instrument remains inert without electricity ; it conditions the form, it does not produce the energy.

It is the same with consciousness and forms ; according to Indian ideas, if you can fabricate the instrument necessary for the manifestation of an energy, that energy can show itself, and what is called consciousness in men is only a part of the universal consciousness which is found everywhere in the universe, and which is translated into human forms.

But they go further : this consciousness is divided into millions of separate parts called *Jîvas* (souls). I do not much care for this word souls - it is quite a theological expression ; they are fragments of life, germs, grains of life, sown in matter. The most subtle form of matter is the first veil of the *Jîva*, an intelligent, conscious being ; this intelligent, conscious being clothes itself with forms of matter of different degrees of subtlety ; these are termed Koshas, a word signifying sheath (the scabbard of a sword, for example), a covering.

There are six of these veils, of these vehicles of consciousness, each coarser than the last. Hence, when consciousness thus veils itself and enters into these vehicles which it has to govern, organize and render fit for its functioning, each vehicle of coarser matter detracts from some of its power. In the first and most subtle matter, it can operate freely ; in the coarser matter some of its powers are lost. Thus consciousness, enveloped in these veils of matter – which are not yet vehicles for consciousness because they cannot act, which are not yet organized –

loses some of its liberty, of its powers, with each additional veil with which it surrounds itself.

It may be asked, why does consciousness clothe itself with these veils? It is because on the highest plane consciousness is vague; it cannot very clearly discern things; it is in the physical body, the vehicle of the coarsest matter, that consciousness can first fabricate the vehicle, of a kind almost perfect, for its manifestation on this plane. Evolution proceeds. Consciousness strives unceasingly to manifest its powers; the *Jîva* works upon the matter, and the vehicles become better and better adapted to its desires.

The man who wishes to evolve more rapidly than by natural processes, adopts methods which have been used for thousands of years, and by which he tries gradually to withdraw consciousness from the coarser material in order that it may function freely in a vehicle of finer material; he endeavours to connect vehicle with vehicle, until he reaches a yet finer vehicle, without ever losing consciousness. In this way it becomes possible to perceive worlds composed of subtler matter, and to observe them, as we observe here, scientifically and directly; and afterwards to remember these observations even whilst wearing the coarsest vehicle, that is to say, the physical body. Such are the ideas of the East.

When man is awake his powers are at their lowest ; when he withdraws from the physical body in the state of sleep, he begins to act in a world composed of somewhat subtler matter. But when he begins to function there, he is not really conscious of himself ; his consciousness is like that of an infant who cannot distinguish between himself and others. But by continuing to function in this way, by repeated experiments, he can attain to self-consciousness on the second plane. If the sleep becomes yet more profound, a yet higher consciousness is revealed, and so on from plane to plane.

Let us note, in passing, that if this theory, proved by many experiments, is true, you have a very lucid interpretation of many of the phenomena of hypnotism and of trance. If it is true that consciousness withdraws from the physical body and functions in a more refined vehicle with enlarged powers, many of these phenomena become intelligible. If, then, you could provisionally accept this theory, it would be possible for you to make some very definite experiments, in order to test the truth of this theory.

I come to another point, and here I am much

afraid of clashing with some scientific opinions. It is believed, by those who hold the Indian theory, that man is not the only conscious being in the universe; they believe that there are many other beings besides man who are intelligent, and who are manifestations of the universal consciousness, and that these beings exist in all the worlds; sometimes they resemble man, at other times they do not resemble him. All around us, in space, that is to say in the other worlds which are in relation with the physical world, are multitudes of intelligent conscious beings, who pursue their lives as we pursue ours; the life is independent, the world is independent, but relations may be established between these worlds.

You doubtless think that you are being transported to the Middle Ages, but these are the Indian ideas of to-day.

It is possible for man when his consciousness begins to function on a supra-conscious plane, to get into relation with these beings and even sometimes to make them obedient to his will, because many of these things are inferior to man.

I have thought it necessary to tell you this because I wish to relate to you two or three experiences which, to me, are unintelligible without this explanation. If you think that this explanation is not valid, find another; for my part I am incapable of doing so.

*
* *

There are in India two great systems of Yoga: the Hâtha-Yoga, that is to say, union by effort; which begins on the physical plane, and does not lead to great heights; and the Râja-Yoga, that is to say, the royal union, an entirely mental system, which does not begin with physical practices, but with mental practices. These then are the two great systems; the Hâtha-Yoga for the body, the Râja-Yoga for the mind, the intelligence.

Those who follow Yoga are called Yogîs. The Hâtha-Yogîs have two aims; one is to secure perfect bodily health and a long extension of life on the earth; the other is to subjugate, for their own advantage, the entities of the other plane, who are not of a very advanced order. It is usually the Hâtha-Yogîs who display phenomena. There is much prejudice in India against other races; they mistrust Westerns and are often reluctant to show them phenomena. I have been able to see a great deal because I have lived among Indians, as an Indian. Indians are very proud; they cannot bear that their ideas, their religion, or their theories, should be laughed at.

The Hâtha-Yogî forces himself to subjugate completely his body and all the functions of his life. Life is called "Prâna," a word usually translated as breath, but it signifies, rather, the aggregation of all the powers of life which are found everywhere. The Hátha-Yogî strives to bring under the control of the human will all the vital functions and to render them absolutely subservient to the will. This is done in two ways; the regulation of the respiration, called "Prânâyâma," a word which means much more than control of the breath, and which signifies control of all the powers of life in the body and even outside the body. The second is "Dhârana," the perfect concentration of attention and of will on a portion of the body. The results obtained by these means are wonderful. The so-called involuntary muscles can be controlled. You may convince yourself by a small experiment on yourself that this is possible. You can easily learn how to move your ear by exercising those muscles which are rudimentary in man. The same can be done with all the muscles of the body. It is possible to entirely stop the heart from beating. The movements first become slower; then the heart ceases to beat and life is as if suspended; the man becomes unconscious on this plane; then little by little, movement is restored until the heart beats regularly. In the same way, the lungs are controlled, always by keeping the attention absolutely fixed on the part that is to be subjugated to the will. One part of the body after another is thus dealt with. These practices last for years.

The Yogî wishes to obtain perfect health; he desires that all the interior of the body should be absolutely clean. The Yogîs make a habit of bathing the interior of their bodies as they do the exterior. They do it sometimes by swallowing through the mouth quantities of water; but they frequently do it also by reversing the peristaltic action of the intestines: they take in water by the lower orifice and eject it by the mouth. I have seen a man who could do that for two or three minutes; he placed himself in water and, after a few moments of these reversed peristaltic movements, he ejected from his mouth what seemed like a fountain of water as long as it was desired that he should do so. This experiment is not beautiful, but it is interesting because it shows the power of the human will when directed upon a portion of the body. It is not then surprising that experiments can be carried out with the human body which seem even less credible.

The result of all these practices is a marvellous state of health, a bodily strength that nothing can break. I have been told – I cannot guarantee this, I am not personally acquainted with an example – that they can sometimes prolong life for a century and a half. Those who have told me this are persons in whom I have the greatest confidence, but, I repeat, I can put forth no proof on this point; what I have observed is the perfect health of these Yogîs.

They attain to complete suppression of the feeling of physical pain. It is thus that a man, whose skin is apparently quite sensitive, can lie on a bed of iron points, and yet appear to feel very comfortable; he feels no pain whatever. Similarly, what would ordinarily be regarded as dreadful suffering is not even felt. A man may have an arm atrophied by holding it raised for years. Imagine the firmness of a will that can do such things. You can understand that with such a will a man can do what he likes with his body.

These life forces in the body which are half conscious, or what you call the Unconscious, do not constitute an elevated order of consciousness ; but they can respond to a higher consciousness, and, in making this response, permit it to control the whole machinery of the body.

This power over the body of suppressing the sensation of pain is found sometimes among those who have not practised the Hâtha-Yoga. One of my friends, of the warrior class, is very fond of tiger hunting ; he is in the habit of going alone into the forest to hunt for tigers ; it is in this way that the warrior class hunt tigers. They do not employ elephants or anything that can protect them in their attack ; they go on foot and quite alone.

One day, however, my friend went tiger hunting with some Englishmen, mounted on their elephants, as is their wont. At the moment when the tiger attacked the elephant, one of the huntsmen lost his presence of mind, his gun went off and the ball lodged in the leg of my friend, who fell. When the surgeon arrived he insisted on putting him under chloroform to extract the ball. My friend refused, and said : "I have never lost consciousness and I do not wish to begin now. Besides, I shall not feel any pain, you may use your knife." The surgeon demurred, saying : "But if you were to make an involuntary movement it might be very dangerous." My friend replied : "I will not move; if I make a single movement I authorise you to use chloroform." The operation was performed; my friend was entirely conscious: he did not make a single movement. What to another would have been horrible torture, was nothing to him.

Afterwards I questioned him on the subject; I thought at first that it was pride of caste that had prevented his showing the least sense of pain. He said to me: "I assure you that I did not feel the least pain. I fixed my consciousness in my head; it was not in my leg; I felt nothing." He was not a Yogî; but he had this power of concentrating his mentality, which is sometimes found among educated Indians. A hereditary physique is transmitted from generation to generation among those who practise Yoga.

The other Hâtha-Yoga which aims at subjugating the beings of another plane, begins always by painful experiments - the *tapas* - such as the one I have just mentioned, namely, holding the arm raised until it becomes absolutely atrophied. They say that it is possible to develop the powers of the consciousness of a plane superior to the physical plane by these extreme austerities (and they do it), and that they can use these powers of the consciousness of the astral plane – that is what they call it – to make use of the inferior entities on that plane. They can thus obtain *apports* of objects without contact ; they can seek what they will, within limits which I will presently indicate ; they can do extraordinary things, which here we should call jugglery, but which are done without apparatus, by will power alone, by the aid, as they say, of these *elementals.* Ten years ago, I saw one of these Yogîs who wished to display some of his powers. He was nearly naked, a consideration of importance when it is a question of the *apports* of objects. He had no sleeves in which he could conceal things. He wore only a little piece of cloth round his loins ; his legs and the upper part of his body – from his waist upwards – were absolutely naked.

He began by one of those feats that can be done here with apparatus, whereas he had only a small table which we ourselves had supplied and a small box with two drawers in it which he allowed us to examine as long and as much as we wished; he had, in addition, an ordinary bottle containing an absolutely clear liquid, like water, but which seemed to me not to be pure water, at least I think not, although I am not sure. We were all seated quite near him ; we could touch the table and assure ourselves that it was not a platform which could conceal trickery.

He first said that he wished to show us some *apports* of objects, and that he had *elementals* under his domination. For a moment, he carefully regarded each of those present. He looked at me fixedly and said: "You must not interrupt me, nor offer any opposition during my operation." I promised, I assured him that I would remain quite passive. I must tell you that I practised Yoga myself before going to India; I think this man was aware of it and clearly perceived that I could oppose his amusements.

He asked three or four of us to entrust him with our watches, and he wrapped them in a handkerchief which we lent him. Then he said to us: "I am going to give this parcel to one of you, that you may take it and throw it into the well." This well was in a little courtyard about fifty yards off. One of our party, a gentleman, took the parcel and went towards the well, when another stopped him, saying: "Perhaps we are the victims of some trickery; let me assure myself that the watches are really in the parcel." The man who said this was a European and thought that this was simply a juggler's trick; he supposed that the Yogî had kept the watches. I do not know where he could have hid them since he was naked. The Yogî got very angry, and said: "Throw the parcel down on the table then." (This anger shows that these men are by no means saints.) One of us opened the parcel; the watches were there. He wrapped them up again, and said: "Give them to Mrs. Besant, who will herself throw them into the well." I took the parcel in my hand, and I went and threw it into the well.

The Yogî was standing by the table. He raised his arms in the air, his hands were empty. He pronounced some words: the watches were in his hands.

Explain that as you like; I confine myself to stating the fact. The man said it was his *elemental* who had fetched the watches out of the well. Perhaps you think these things are quite impossible; they will seem to you incredible if you have not been present at spiritistic seances where just the same kind of things are done, where objects are brought as *apports* without contact. The handkerchief which was wrapped round the watches was quite wet.

The man next suggested cutting off the head of a bird, assuring us that it would not hurt it. I did not wish to witness such a painful experiment. I only wished to see what could be seen without horror. He assured us that he could perform this experiment; but I think that this must be produced by collective hallucination, whilst I do not think that in the experiment with the watches there was any hallucination. And assuredly, there was no hallucination in the following experiment: –

"Ask me," he said, "to bring something to you; my elemental will bring it in a box." Someone enquired if he could cause objects to be brought from a distant country. "I can if they are in India," he replied, "but it is not possible if the sea must be crossed." Here, therefore, was a limit to his powers. Someone then said to him: "At a distance of a hundred miles from here there is a town where a kind of sweet is made that is found nowhere else in India. Will you bring us some of these sweets?" The man stood in the midst of our circle in full light, it was morning. He opened the box and began emptying it with both hands; he threw some sweets on the table and soon made a pile of them much higher than the box. He said that it was his elemental who had brought them. They were really the sweets asked for; we distributed them among the neighbouring children, who ate them with much pleasure.

These are but a few of those experiments which are very difficult for Western minds to comprehend, but very easy for an Indian to explain by his theory of consciousness and of the elemental. You might try to make these experiments; perhaps you may succeed, perhaps you may not succeed.

I have been told of an experiment which I have not seen; it is very well known, it is that of the basket and the little child; perhaps I should say that I have seen it once, but I am convinced that it was jugglery and not the effect of Hâtha-Yoga. One of my friends, an officer in the English army, told me that he had seen this experiment performed in the courtyard of his own house. He stood on one side of the basket and a brother officer stood on the other; they saw the child who was put into the basket; they themselves tied it with cords; they did not move away from the basket, and they did not lose sight of it for a single moment. The man was in front of the basket; he began singing in a low voice a strange refrain, which lasted for ten minutes. After that he proceeded in the usual way.* When that was over, and after a great quantity of blood had

* That is to say, he pierced the basket repeatedly, in every direction, with a sword.

been seen issuing from the basket, the child appeared amidst the crowd of onlookers safe and sound.

I can only explain that as a collective hallucination. There are things which can be achieved by those who have a more extended knowledge of nature ; but on the physical plane, to stick a sword into the body of a child, to shed its blood abundantly and to cause the child afterwards to reappear is impossible, it is contrary to known physical laws. It was his strange chant that induced the collective hallucination. They have very strange chants which produce marvellous effects on the brain ; it is thus that they hypnotise a crowd, which sees only what the hypnotiser wills shall be seen.

This experiment, therefore, is not interesting to me ; it is fairly easy ; it consists in the knowledge of a succession of sounds that hypnotise. This is the secret which is generally in the possession of some family, and is transmitted from generation to generation. Moreover, each family can perform only one kind of experiment, one sort of hallucination.

These Yogîs can put themselves into auto-hypnotic trances with great facility ; but these trances, when they come out of them, do not seem to leave them with any fresh knowledge ;

the trance is therefore absolutely useless. I have seen a Yogî who was always in a state of absolute unconsciousness on the physical plane ; his disciples took care of him, and fed him ; he was like an idiot and had nothing to teach.

These men have developed the power of hypnotising themselves ; but they have not developed the capacity of possessing consciousness on a superior plane which can be transmitted to the brain.

The Yogîs can predict the exact hour of their death, that is to say they can choose this hour. I know one who said : "I will die to-day at five o'clock." His disciples were with him, and at five o'clock exactly he died. They are able to quit their bodies either in a trance, from which they can return, or in death, from which they do not return. They generally die in this way, choosing the exact hour at which they wish to quit their bodies.

*
* *

The other method, the Râja-Yoga, is quite different. There are in Yoga eight successive degrees : Yâma, Niyâma, Asana, Prânâyâma, Pratyahāra, Dhārana, Dhyâna, Samâdhi ; in Hâtha-Yoga one begins with the third degree, that is to say with Asana, the posture. The posture in which the body is held is of great importance in relation to the vital currents. Some of these postures are very difficult, some are quite easy. The Hâtha-Yogî assumes very difficult and painful postures. The Râja-Yogî does not as a rule assume difficult postures for the body, but chooses, rather, the easy ones. Patanjali says:* "An easy and pleasant posture."

In the Râja-Yoga one begins by the first two degrees, that is to say, by the moral; purification is needful. This is not necessary for the Hâtha-Yoga. The first step, Yâma, is negative purification: that is, complete abstinence from all that is evil; not a single creature must be injured, a man must live in perfect charity towards all. The second step is Niyâma, that is, positive purification: the practice of the virtues helpful to humanity. Without this there is no Râja-Yoga; these two rungs of the ladder are absolutely necessary. Then a bodily posture must be chosen (Asana) which can be maintained for a long time without fatigue; it is only necessary to keep the back, the throat and the head in a straight line, that

* *The Yoga-Sutra of Patanjali*, translated by Manilal Dvivedi, Bombay Tookárám Tátyá.

is to say, that the vertebrate column should be quite straight in order that the currents may pass without obstacle. The head must not be turned to the right or to the left ; to keep the body quite straight is the only position necessary for the Râja-Yoga.

After this comes Prânâyâma, that is to say, the control of the powers of life in the body. Then the Pratyhâra, in which the mind is not concentrated upon one part of the body; but all the mental faculties are gathered together. They are diverted from external objects in order to observe nothing of the environment in which one is placed. All the avenues of sense are closed. At first they are usually closed in a physical manner ; there is a way of placing the fingers so as to close at one and the same time the nostrils, the eyes and the ears. But when concentration has been developed, it is no longer necessary to employ these means; the senses cease to function. This is attained simply by mental effort, a method the very opposite of that employed in hypnotism, where the senses are fortified by turning a mirror, for example. This is called collecting the forces, turning the mentality within ; there is then perfect concentration (Dhârana), not upon one part of the body, but upon an idea ; there is a mental image, an image which one must strive to make very clear, very precise.

These are the inferior degrees: their object is to liberate consciousness from the body. When the senses no longer function, when the exterior environment has disappeared, when one has become insensible to external contact, consciousness begins to function in a more subtle vehicle belonging to the Beyond; it truly functions; this is what is called in the West the supraliminal or supra-consciousness. The superior consciousness must work in the world beyond, and make observations; this is termed Dhyâna, meditation.

If a yet higher plane is reached, one which is called Samâdhi (a supra-consciousness which is conscious of itself) it is possible on returning thence to the body to use the physical brain to remember the observations which have been made on other planes.

Such is the conception of the Ràja-Yoga, a development more and more intense of the mental powers, complete insensibility to the senses, but perfect interior consciousness.

In this condition the Yogî can vacate his body consciously without losing consciousness, and having left his body can perceive it distinctly lying there as an exterior object beside him.

Then the conscious being, who is thus able to regard his body like a cast-off garment, can rise from one sphere to another, make his observations, fix them on the memory, and impress them on the brain, so that they will persist when he returns to the body.

The proof that the body has been really vacated is that knowledge may thus be acquired which is not possessed on the physical plane; and different persons may compare their experiences. Their observations will not be entirely identical, because the play of personality always enters as a factor into the experience, but it is possible to make observations of so precise a kind, that it may easily be perceived that the slight variations in detail are due to differences in the observers, and not to differences in the objects observed.

If you interrogate a dozen persons who have passed at the same hour down the same street, they will tell you very different things; because as the mentality of each person differs, their observations are different. Nevertheless, by their several accounts, even though different, you will have no difficulty in recognising the street of which they speak.

Thus many persons have been able to observe the same objects in another world and to register their observations when they have returned to the physical body.

If this is possible, it explains many phenomena noted in psychical research. We can understand why consciousness in a state of trance is something much keener, and has a much more extended knowledge than in the waking state. If, however, we can have this personal experience of the supra-consciousness, and return to the physical body, we possess satisfying proof and invincible certitude of the persistence of consciousness apart from the physical body.

May I suggest that modern psychologists should make very careful study of the class of experiences called religious;* the religious consciousness of monks and nuns and saints is still consciousness. It may be said that it is a deformed consciousness; but sometimes a deformed consciousness exhibits facts of great value.

In India they tell us that the brain is destroyed if it is not trained in a certain way before it is allowed to receive the impressions of the supra-consciousness. The brain, indeed, cannot bear, without risk, these intense, rapid vibrations of the supra-consciousness; and, before trying

* See *Varieties of Religious Experience*, by Prof. William James.

these experiments, it is necessary to exercise the brain by thinking the highest and sublimest thoughts. If by intense emotion a man throws himself into the other world, when he returns to his body, hysteria is sure to follow those vibrations ; the brain cannot endure these vibrations without preparation, but they can be endured by means of Yoga practices. It has often been stated that those who have given themselves up to these experiments in monasteries or elsewhere, have suffered from lack of sleep or from nervous troubles suggestive of hysteria. That is quite true and I do not wish to deny it ; but I say that this is not inevitable. If we proceed step by step, if a strong will creates a suitable condition of the nervous system, the brain may become keener, and at the same time remain absolutely healthy ; then you have the Yogî instead of the hysteric.

In conclusion : I have sketched a theory which you can study ; you can make experiments in order to discover whether this theory does, or does not, explain the problems that modern psychology cannot solve. The latter collects numbers of facts, but it cannot always explain them. It appeals to the Unconsciousness : but there is not only one Unconsciousness : there is the unconsciousness which is derived from the past, that is, the sub-consciousness; the Hâtha-Yogî makes this, too, to become conscious and governs all the movements of the body. Then there is the supra-consciousness, which is the Consciousness of the future, for which the physical body is not yet sufficiently evolved. Therefore experiments with this supra-consciousness present many dangers. It will however be the normal consciousness of the future. Human evolution is not finished; man is still very imperfect; it is possible to put pressure on the body, to make it work in such a way as to hasten the normal advances of evolution. If this is done with precaution, with knowledge, with the help of those who know the way, it is possible to walk along this path without danger, without injuring the body, without becoming a hysteric, without nervous degeneration, and it is just this idea that I have desired to lay before you in this paper.

Men and Animals.

From "Bibby's Quarterly," Autumn. 1903.

THE relation of man to the lower animals is but very partially understood, chiefly because animals are generally regarded as "having no souls," and hence as being divided from man by an insuperable gulf. In Italy this idea has been carried so far that even cruelty is excused, under the plea "*Non e Christiano,*" "It is not a Christian," as though the absence of a future life justified the making of the present life miserable! But even among kindly-hearted people there is a very general idea that animals are merely an appendage of man, and that, as it is often phrased, "God made animals for man's use." Hence the animal is regarded only in the light of its usefulness to man, and to consider the welfare and evolution of the animal as a separate being would, to most people, savour of the absurd.

Yet it is not absurd if the animal, like man, should be an evolving creature, if the animal should in some sense have a "soul." Now, in the animal we find maternal affection, capacity of love, fear of pain, and dawning intelligence, and in some we see great courage, endurance, fidelity and devotion to a master. Great as are the differences between these and the corresponding qualities in a civilised man, they are differences in degree rather than in kind, and a better moral character may be found in a domesticated dog than in a low type of savage.

A brave, loving dog, faithful to death, would seem to be more worthy of immortality than a blood-thirsty, cruel, treacherous savage. Yet ordinary orthodoxy dooms the one to extinction and awards immortality to the other.

Now, it is true that there is one important difference between an animal and a man ; both are vivified by an immortal Spirit, whose powers are more or less unfolded and active ; but the bridge between the immortal spirit and the perishing body, that which is sometimes called "the soul," the intelligent, self-conscious "I," is present in the man, even in the most brutal savage, and is normally absent in the animal.

Take a flock of sheep, a herd of cattle, any group of similar animals, wild or domesticated, and a marked similarity of thought, feeling and action may be observed among them. They are largely guided by instincts, which they share in common, and comparatively little by individual reasoning; it is as though there were "a common soul" guiding them all.

But when one of the higher animals comes into close relations with men - such an animal as a dog or a cat - a gradual change is visible to the close observer. If the animal be a favourable specimen of its class, and be strongly devoted to its owner, it will gradually separate itself off from its kind, and begin to show marks of individuality; it will evolve strong likes and dislikes, will follow ways of its own, will manifest ever-increasing powers of reasoning, and anyone who can use clairvoyant vision will see that a change has taken place in the superphysical bodies of such an animal.

Now, a man, however undeveloped, however savage, shews an astral body, a mental body, and a causal body, with the spirit brooding over and vivifying all. But an animal shows an astral body, a vague cloud representing an embryonic mental body and the over-brooding spirit: the causal body, that which makes possible the self-conscious "I," is absent. Herein lies the difference between the animal and man, between the noblest ordinary animal and the most brutal savage.

But when a highly-developed animal becomes intensely devoted to some human being, and clings to him with passionate and unwavering fidelity, the play of the human self-conscious intelligence stimulates the dawning intelligence and quickens the unfolding of the spirit in the animal, and at last, as it were, a flash, like an electric spark, springs across the gulf between the over-brooding spirit and the embryonic mental body, a bridge of light spans the gulf, the causal body is formed, the " soul " is born. Henceforth that animal is separated off from its kind, and has completed the term of animal evolution. Its death will be followed by an immense period of rest and inner growth, and it will, long ages hence, be borne into some future humanity, to begin the long course of human evolution.

These individualised animals are, indeed, rare exceptions, but all animals are treading the path which leads to individualisation, and their progress is hastened or retarded by the human beings with whom they come into contact. The dog, the cat, and the horse are the three animals capable of profiting most by association with man, and their progress in the animal kingdom may be much quickened by the wise, firm, and sympathetic training given to them by their elder brothers, men. Even when they may not reach the point of individualisation, they may be led up near to it, and a link is made between them and their masters which, in the future, will be a source of benefit and happiness to both.

The practical difference which the adoption of this view of animals would cause in the relationship between them and men would not be the relinquishment of their services, nor the loss of their utility. They would be used as much as they are now, but the treatment of them would be always kind, considerate, firm and judicious. The training of the animal would be regarded as useful to the animal as well as to the man ; hasty and unnecessary blows and savage language would be avoided, and harsh punishment of horse or dog would be regarded as showing the owner's incapacity to train and educate aright. All cruel methods of breaking would be abandoned, trust and confidence on the part of the animal would be encouraged, and we should hear much less of "incurably vicious horses" - vice which is mostly the result of human stupidity and cruelty.

Man must gradually learn to regard himself as the divinely appointed ruler of the animal world, using his great powers to raise and train his subjects, not to crush and terrify them. He must cease to look on them as existing only for his use and comfort, and regard them as his infant brethren in the divine family, knowing that he is the representative to them of the Divine Being, to whom he must answer for the exercise of the kingship placed in his hands.

9 780766 191273